MATHS
NOW!
• • • • • • •
GCSE

higher

1

MATHS NOW!

GCSE

higher 1

MATHS NOW!
National Writing Group

JOHN MURRAY

Acknowledgements

The authors and publishers would like to thank all the teachers, schools and advisers who evaluated *Maths Now!* and whose comments contributed so much to this final version.

Particular thanks go to: David Bullock, Frodsham High School, Warrington; Philip Chaffé, King Edward VI School, Lichfield; John D. Collins, Education Consultant and Inspector of Schools; Patrick Gallagher, Convent of Jesus and Mary RC High School, London; Ian Gregory, Blakeston Community School, Stockton-on-Tees; David McLaren, Consultant in Mathematical Education; Peter Marks, Ilfracombe College, Devon; Kevin Pankhurst, Pilton Community College, Barnstaple; Kim O'Driscoll-Tole, University of Strathclyde; Mrs T. Stephens, Bedwas School, Gwent.

Photo acknowledgements

Cover: John Townson/Creation; **p.6** John Townson/Creation; **p.11** Mary Evans Picture Library; **p.30** © John P. Stevens/Ancient Art & Architecture Collection; **p.78** *left* AKG London/Erich Lessing, *right* AKG London; **p.110** © Mitchell Coster/Axiom; **p.118** Dr Fred Espenak/Science Photo Library; **p.223** Werner Forman Archive; **p.253** © Marilyn Bridges/Corbis; **p.260** John Townson/Creation; **p.285** John Townson/Creation; **p.286** © R. Sheridan/Ancient Art & Architecture Collection.

© R. C. Solomon/MATHS NOW! National Writing Group 2001

First published in 2001
by John Murray (Publishers) Ltd
50 Albemarle Street
London W1S 4BD

Layouts by Stephen Rowling/springworks
Artwork by Oxford Designers & Illustrators
Spreadsheet screen shots by Ric Pimentel
Cover design by John Townson/Creation

Typeset in 10/12 Times by Wearset, Boldon, Tyne and Wear
Printed and bound in Great Britain by Butler and Tanner, Frome and London

A catalogue entry for this title is available from the British Library

ISBN 0 7195 7445 5
Teacher's Resource Book GCSE Higher 1 0 7195 7446 3

Contents

Introduction

Now you have put Key Stage 3 behind you, and have started on Key Stage 4. At the end of the second year of Key Stage 4 you will take the GCSE exam.

For some of you this exam will mark the end of your mathematics at school, and it will be the evidence of your mathematics education that you take out into the world. This book and its sequel provide complete coverage of everything you need for the exam and, we hope, an insight into the extent and richness of mathematics.

Some of you will be considering continuing with mathematics after GCSE. This book and its sequel contain extra exercises of preparation for AS and A level Mathematics. Mathematics is an open-ended subject – there are always new questions to be posed and new answers to be found.

Symbols

The symbols used in the Student's Book are as follows:

Use a calculator

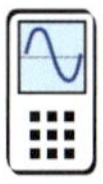

Use a graphics calculator

Do not use a calculator

This is a particularly challenging question/exercise

Ma1

1 Indices

The four basic operations of arithmetic are addition, subtraction, multiplication and division. A fifth operation is taking powers. You have already met the expression x^n, for $x \neq 0$, when n is an integer. Here we extend the definition to fractional n. First, let's have a reminder of the definitions and rules for integer values of n.

To find the nth power of a number x, multiply together n of the number x.

$$x^n = \underbrace{x \times x \times \ldots \times x}_{n \text{ factors}}$$

So, for example,

$$4^3 = 4 \times 4 \times 4$$

The expression x^n is a **power**. The number n is the **index** (plural indices) and the number x is the **base**. So, for the power 4^3, the index is 3 and the base is 4.

When powers of the same base are multiplied, the indices are added.

$$x^m \times x^n = x^{m+n}$$

So, for example, $3^5 \times 3^6 = 3^{11}$.

When powers of the same base are divided, the indices are subtracted.

$$x^m \div x^n = x^{m-n}$$

So, for example, $5^9 \div 5^3 = 5^6$.

Taking powers has priority over the other operations. For example, 2×3^2 means that you **square** 3 and then multiply by 2. The result is 18. If you want to do the multiplication first, use brackets.

$$2 \times 3^2 = 2 \times 9 = 18 \qquad (2 \times 3)^2 = 6^2 = 36$$

Exercise 1.1

Evaluate the expressions in questions 1–16.

1 2^4 **2** 3^3 **3** 5^4 **4** 7^2
5 6^3 **6** 10^4 **7** $(\frac{1}{2})^4$ **8** $(\frac{1}{3})^3$
9 0.4^3 **10** $(1\frac{1}{2})^3$ **11** 5×2^2 and $(5 \times 2)^2$ **12** $8 \div 4^3$ and $(8 \div 4)^3$
13 $8 - 2^3$ and $(8 - 2)^3$ **14** $2^3 + 1$ and $(2 + 1)^3$ **15** $3^3 \times 2$ and $(3 \times 2)^3$ **16** $20 \div 10^2$ and $(20 \div 10)^2$

Write the expressions in questions 17–26 as powers of a single base.

17 $x^3 \times x^7$ **18** $y^{10} \div y^3$
19 $2^m \times 2^n$ **20** $3^x \div 3^y$
21 $2^a \times 2^b \times 2^c$ **22** $x^3 \times x^5 \times x^6$
23 $y^5 \times y^8 \div y^3$ **24** $3^x \times 3^y \div 3^z$
25 $\dfrac{5^a \times 5^b}{5^c}$ **26** $\dfrac{x^8}{x^3 \times x^2}$

Zero and negative indices

The definition and the rules make sense if the indices are positive whole numbers. We can extend the definition so that it holds for indices which are zero or negative.

provided $x \neq 0$

$$x^0 = 1$$

$$x^{-n} = \frac{1}{x^n}$$

Justification

Suppose $x \neq 0$. Then consider x^1 divided by itself, using the rule for division of powers.

$$x^1 \div x^1 = x^{1-1} = x^0$$

But $x^1 \div x^1 = x \div x = 1$. Hence $x^0 = 1$.

Suppose $x \neq 0$. Consider x^0 divided by x^n, using the rule for division of powers.

$$x^0 \div x^n = x^{0-n} = x^{-n}$$

But x^0 is equal to 1. Hence $x^{-n} = 1 \div x^n$, that is $\dfrac{1}{x^n}$.

Exercise 1.2

Evaluate these expressions.

1 3^0
2 $k^0, k \neq 0$
3 2^{-1}
4 5^{-1}
5 10^{-2}
6 2^{-4}
7 1^{-3}
8 $\left(\frac{1}{2}\right)^{-1}$
9 $\left(\frac{1}{3}\right)^{-2}$
10 $\left(1\frac{1}{2}\right)^{-2}$
11 $\left(1\frac{1}{3}\right)^{-2}$
12 $\left(\frac{2}{3}\right)^{-3}$

The rules for multiplying and dividing powers hold, even when the indices are zero or negative.

Example   Simplify $x^{10} \div x^{-8}$.

Use the rule that the indices are subtracted.

$$10 - (-8) = 10 + 8 = 18$$

Hence $x^{10} \div x^{-8} = x^{18}$.

Exercise 1.3

Simplify the expressions in questions 1–25. In all cases the variables are non-zero.

1 $x^6 \div x^{10}$
2 $k^2 \div k^9$
3 $y^3 \times y^{-8}$
4 $a^4 \times a^{-7}$
5 $k^{-7} \times k^{-5}$
6 $d^{-4} \times d^{-7}$
7 $n^{-3} \div n^6$
8 $m^{-2} \div m^8$
9 $y^{-2} \div y^{-8}$
10 $z^{-3} \div z^{-5}$
11 $m^9 \div m^{-8}$
12 $3^x \times 3^{-y}$
13 $2^a \times 2^{-b}$
14 $7^{-m} \times 7^{-n}$
15 $2^{-x} \times 2^{-y}$
16 $5^a \div 5^{-b}$
17 $3^x \div 3^{-y}$
18 $2^{-q} \div 2^p$
19 $5^{-a} \div 5^b$
20 $3^{-x} \div 3^{-y}$
21 $2^{-p} \div 2^{-q}$
22 $\dfrac{2^{-x} \times 2^{-y}}{2^{-z}}$
23 $\dfrac{3^a \times 3^b}{3^{-c}}$
24 $\dfrac{5^{-m}}{5^{-n} \times 5^p}$
25 $\dfrac{2^a \times 2^b}{2^c \times 2^{-d}}$

Powers of products and quotients, and powers of powers

Powers of products and quotients

Suppose you have a product (two or more terms multiplied together) and raise it to a power. The result is a power of a product. For example, $(xy)^3$ is the third power of the product xy.

Similarly, suppose you have a quotient (one term divided by another) and raise it to a power. The result is a power of a quotient. For example, $\left(\dfrac{a}{b}\right)^4$ is the fourth power of the quotient $\dfrac{a}{b}$.

On page 1 we pointed out that taking a power has priority over other operations. If you want to do another operation first, use brackets. The brackets are necessary in an expression like $(xy)^3$.

$$2 \times 3^2 = 2 \times 9 = 18 \quad \text{but} \quad (2 \times 3)^2 = 6^2 = 36$$

We can expand a product or a quotient raised to a power – just take the power of the individual terms.

$$(2x)^3 = 2^3 \times x^3 = 8x^3 \qquad \left(\frac{3}{b}\right)^4 = \frac{81}{b^4}$$

● *In general,*

$$(xy)^n = x^n y^n \qquad \left(\frac{a}{b}\right)^n = \frac{a^n}{b^n}$$

> $(x+y)^n$ is not equal to $x^n + y^n$.

Justification

Consider $(xy)^n$. This consists of n copies of xy, multiplied together.

$$(xy)^n = xy \times xy \times xy \times xy \times \ldots \times xy$$

This is n xs and n ys, multiplied together.

$$(xy)^n = x^n \times y^n$$

Powers of powers

When a power is raised to another power, the result is a power of a power. For example, $(2^3)^4$ is the fourth power of the power 2^3. To evaluate a power of a power, multiply the indices together.

> This is the same as x^{mn}.

$$(x^n)^m = x^{nm}$$

Justification

Consider $(x^n)^m$. This is m copies of x^n, multiplied together.

$$(x^n)^m = x^n \times x^n \times x^n \times \ldots \times x^n$$

The total number of factors of x is $n + n + n + \ldots + n$, which is nm.

$$(x^n)^m = x^{nm}$$

Example   Write $25x^4$ as a single power using brackets.

Note that $25 = 5^2$, and that $x^4 = (x^2)^2$. Both terms are squares, hence we can put them inside brackets.

$$25x^4 = (5x^2)^2$$

Exercise 1.4

In questions 1–16, expand the brackets.

1 $(3n)^3$ **2** $(5n)^2$ **3** $(ax)^3$ **4** $(2by)^4$

5 $(\frac{1}{2}x)^2$ **6** $(\frac{1}{3}x)^3$ **7** $\left(\dfrac{x}{2}\right)^4$ **8** $\left(\dfrac{k}{3}\right)^3$

9 $\left(\dfrac{a}{b}\right)^5$ **10** $\left(\dfrac{2y}{3}\right)^3$ **11** $(2^3)^2$ **12** $(x^3)^2$

13 $(a^3)^5$ **14** $(k^3)^{-2}$ **15** $(n^{-2})^3$ **16** $(m^{-5})^{-2}$

In questions 17–31, write the expression as a single power using brackets.

17 a^3b^3 **18** $8x^3$ **19** $16y^{-4}$ **20** $100k^{-2}$

21 $\dfrac{x^3}{y^3}$ **22** $\dfrac{a^4}{16}$ **23** $\dfrac{k^2}{49}$ **24** $\dfrac{100}{x^2}$

25 $4x^4$ **26** $8a^6$ **27** $\frac{1}{4}k^2$ **28** $16x^2$

29 $81a^2$ **30** $\dfrac{x^4}{9}$ **31** $\dfrac{49}{y^6}$

In questions 32–36, expand and simplify the expressions.

32 $(5x)^2 + (3x)^2$ **33** $(2z)^3 + (4z)^3$ **34** $(3a)^2 - (2a)^2$ **35** $(3xy)^3 - (2xy)^3$
36 $(\frac{1}{2}ab)^2 + (\frac{1}{3}ab)^2$

37 Write out the justification for the rule $\left(\dfrac{a}{b}\right)^n = \dfrac{a^n}{b^n}$.

We cannot simplify $2^3 \times 3^4$. But we can simplify $2^3 \times 4^4$, by writing 4 as a power of 2. $4 = 2^2$, hence $2^3 \times 4^4 = 2^3 \times (2^2)^4 = 2^3 \times 2^8 = 2^{11}$.

Example Write as a power of 2: $\dfrac{2^3 \times 4^2}{8^4}$.

Both 4 and 8 are powers of 2: $4 = 2^2$ and $8 = 2^3$.

$$\frac{2^3 \times 4^2}{8^4} = \frac{2^3 \times (2^2)^2}{(2^3)^4} = \frac{2^3 \times 2^4}{2^{12}}$$

Now add and subtract the indices: $3 + 4 - 12 = -5$.

$$\frac{2^3 \times 4^2}{8^4} = 2^{-5}$$

It may help to write out the expression in full.
$$\frac{2^3 \times (2^2)^2}{(2^3)^4} = \frac{(2 \times 2 \times 2) \times (2 \times 2) \times (2 \times 2)}{(2 \times 2 \times 2) \times (2 \times 2 \times 2) \times (2 \times 2 \times 2) \times (2 \times 2 \times 2)} = \frac{2^7}{2^{12}} = 2^{-5}$$

Exercise 1.5

1 Write 9 as a power of 3. Hence write $3^3 \times 9^2$ as a power of 3.
2 Write 49 as a power of 7. Hence write $49^3 \times 7^5$ as a power of 7.
3 Write 25 and 125 as powers of 5. Hence write $125^3 \div 25^2$ as a power of 5.

In questions 4–14, write the expressions as powers of a single number.

4 $2^3 \times 4^2 \times 8^2$

5 $9^2 \times 3^3 \div 27^5$

6 $1000 \times 100^4 \div 10^5$

7 $2^m \times 4^n$

8 $3^a \div 9^b$

9 $4^3 \times 8^2$

10 $9^x \times 27^y$

11 $\dfrac{125^4}{5^2 \times 25^3}$

12 $\dfrac{1000^3 \times 10^4}{100^2}$

13 $3^x \times 9^y \times 27^z$

14 $\dfrac{16^a \times 2^b}{4^c \times 8^d}$

Fractional indices

We are able to define powers with negative indices, such as 5^{-3}, even though it doesn't make sense to say 'minus three fives multiplied together'. What if the index is a fraction, as in $5^{\frac{1}{2}}$? It doesn't make sense to say 'half a five multiplied together' either.

We can define fractional indices, so that they agree with the rules for multiplying and dividing powers. Consider $5^{\frac{1}{2}}$. Multiply this number by itself, that is, find the square of $5^{\frac{1}{2}}$. Use the rule for powers of powers.

$$(5^{\frac{1}{2}})^2 = 5^{\frac{1}{2} \times 2} = 5^1 = 5$$

So when $5^{\frac{1}{2}}$ is squared, the result is 5. So $5^{\frac{1}{2}}$ must be the **square root** of 5.

$$5^{\frac{1}{2}} = \sqrt{5}$$

Try with another number. Try squaring $20^{\frac{1}{2}}$.

$$(20^{\frac{1}{2}})^2 = 20^{\frac{1}{2} \times 2} = 20^1 = 20$$

So $20^{\frac{1}{2}}$ is the square root of 20.

$$20^{\frac{1}{2}} = \sqrt{20}$$

● *For any number which isn't negative, the $\frac{1}{2}$th power is the square root of the number.*

$$if \ x \geq 0, \ x^{\frac{1}{2}} = \sqrt{x}$$

Example   Find $\sqrt{0.64}$.

Write as $0.64^{\frac{1}{2}} = \left(\dfrac{64}{100} \right)^{\frac{1}{2}} = \dfrac{64^{\frac{1}{2}}}{100^{\frac{1}{2}}} = \dfrac{8}{10}$

Hence $\sqrt{0.64} = 0.8$.

Exercise 1.6

In questions 1–8, evaluate the expressions. Do not use a calculator.

1 $100^{\frac{1}{2}}$ **2** $49^{\frac{1}{2}}$ **3** $64^{\frac{1}{2}}$ **4** $0.25^{\frac{1}{2}}$

5 $0.01^{\frac{1}{2}}$ **6** $0.49^{\frac{1}{2}}$ **7** $(1\frac{7}{9})^{\frac{1}{2}}$ **8** $(2\frac{1}{4})^{\frac{1}{2}}$

In questions 9–13, write the expressions in power form.

9 $\sqrt{5}$ **10** $\sqrt{0.002}$ **11** $\sqrt{x}$ **12** $\sqrt{k}$ **13** $\sqrt{(5x)}$

Simplify the expressions in questions 14–18.

14 $\sqrt{(x^2)}$ **15** $(y^4)^{\frac{1}{2}}$ **16** $(a^{\frac{1}{2}})^6$ **17** $(x^2y^2)^{\frac{1}{2}}$ **18** $(a^{\frac{1}{2}}b^{\frac{1}{2}})^2$

19 Show that $\sqrt{12} = 2\sqrt{3}$. Confirm this by using a calculator.

Exercise 1.7

You now have three ways to find a square root using a scientific calculator.

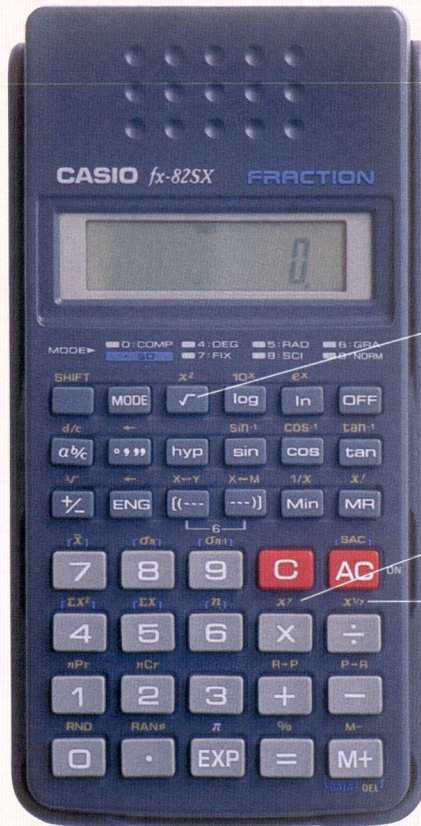

Use the square root button, labelled $\sqrt{\ }$

Use the x^y button, taking y equal to 0.5

Use the $x^{\frac{1}{y}}$ button, taking y equal to 2

Find $\sqrt{5}$ in these three different ways, checking that the answers are the same.

The $x^{\frac{1}{y}}$ button is sometimes called $\sqrt[x]{y}$. In this case, take x equal to 2.

What about other fractional powers, such as $5^{\frac{1}{3}}$? Use the same method as for $5^{\frac{1}{2}}$, by finding the **cube** of $5^{\frac{1}{3}}$.

$$(5^{\frac{1}{3}})^3 = 5^{\frac{1}{3} \times 3} = 5^1 = 5$$

So when $5^{\frac{1}{3}}$ is cubed, the result is 5. It follows that $5^{\frac{1}{3}}$ is the cube root of 5.

$$5^{\frac{1}{3}} = \sqrt[3]{5}$$

What about $7^{\frac{1}{4}}$? Try the fourth power of $7^{\frac{1}{4}}$.

$$(7^{\frac{1}{4}})^4 = 7^{\frac{1}{4}\times 4} = 7^1 = 7$$

So the fourth power of $7^{\frac{1}{4}}$ is 7 itself. It follows that $7^{\frac{1}{4}}$ is the fourth root of 7.

$$7^{\frac{1}{4}} = \sqrt[4]{7}$$

● *In general, for any non-negative number x, the $\frac{1}{n}$-th power of x is the nth root of x.*

$$x^{\frac{1}{n}} = \sqrt[n]{x} \quad provided\ x \geqslant 0$$

Exercise 1.8

In questions 1–5, evaluate the expressions. Do not use a calculator.

1 $27^{\frac{1}{3}}$ **2** $1000^{\frac{1}{3}}$ **3** $32^{\frac{1}{5}}$ **4** $10000^{\frac{1}{4}}$ **5** $(\frac{1}{8})^{\frac{1}{3}}$

In questions 6–12, write the expressions in terms of roots.

6 $x^{\frac{1}{3}}$ **7** $y^{\frac{1}{5}}$ **8** $a^{\frac{1}{10}}$ **9** $10^{0.2}$
10 $8^{0.125}$ **11** $2^{\frac{1}{x}}$ **12** $a^{\frac{1}{n}}$

Using a calculator

A scientific calculator has a button for finding nth roots. It might be labelled $x^{\frac{1}{y}}$ or $\sqrt[x]{y}$.

The sequence to find $32^{\frac{1}{5}}$, that is, $\sqrt[5]{32}$, might be

| 32 | $x^{\frac{1}{y}}$ | 5 | = |

The answer 2 should appear. Try it.

Exercise 1.9

Go through questions 1–5 of exercise 1.8, this time using the $x^{\frac{1}{y}}$ (or $\sqrt[x]{y}$) button on a calculator. Check that your answers are the same as before.

What about a power like $10^{\frac{2}{3}}$? We can use the rule about powers of powers.

$$10^{\frac{2}{3}} = 10^{\frac{1}{3}\times 2} = (10^{\frac{1}{3}})^2 = (\sqrt[3]{10})^2$$

So, to find $10^{\frac{2}{3}}$, take the cube root of 10 and then square it. Similarly, $6^{\frac{3}{4}}$ is the fourth root of 6, cubed.

$$6^{\frac{3}{4}} = (\sqrt[4]{6})^3$$

● *In general, $x^{\frac{n}{m}}$ is the mth root of x, taken to the nth power.*

$$x^{\frac{n}{m}} = (\sqrt[m]{x})^n$$

Note. We can do the operations in the opposite order. We can take the nth power of x, then take the mth root.

$$x^{\frac{n}{m}} = \sqrt[m]{x^n}$$

Examples Find $8^{\frac{2}{3}}$.

This is the cube root of 8, squared. The cube root of 8 is 2, and 2 squared is 4.

$$8^{\frac{2}{3}} = (\sqrt[3]{8})^2 = 2^2 = 4$$

Alternatively, if we do the operations in the other order, $8^{\frac{2}{3}} = \sqrt[3]{8^2}$. Hence

$$8^{\frac{2}{3}} = \sqrt[3]{64} = 4$$

This method involves larger numbers.

Find $49^{-\frac{1}{2}}$.

Write the index as $-1 \times \frac{1}{2}$, and then use the rule.

$$49^{-\frac{1}{2}} = 49^{-1 \times \frac{1}{2}}$$
$$= (49^{\frac{1}{2}})^{-1} = 7^{-1} = \tfrac{1}{7}$$

$49^{\frac{1}{2}} = \sqrt{49} = 7$

Exercise 1.10

Evaluate these expressions. Do not use a calculator.

1 $4^{\frac{3}{2}}$
2 $1000^{\frac{2}{3}}$
3 $16^{\frac{3}{4}}$
4 $4^{-\frac{1}{2}}$
5 $(\frac{1}{9})^{-\frac{1}{2}}$
6 $32^{0.6}$ (Hint: $0.6 = \frac{3}{5}$)
7 $81^{\frac{3}{4}}$
8 $(\frac{1}{9})^{\frac{3}{2}}$
9 $(\frac{1}{8})^{-1\frac{1}{3}}$
10 $1024^{0.3}$
11 $16^{0.75}$
12 $0.125^{\frac{1}{3}}$
13 $(3\frac{3}{8})^{\frac{2}{3}}$
14 $2.25^{2.5}$

The rules of powers apply when the index is fractional.

Example Find $3^{2\frac{1}{2}} \times 3^{-1\frac{1}{2}}$.

Add the indices. $2\frac{1}{2} + (-1\frac{1}{2}) = 1$.

$$3^{2\frac{1}{2}} \times 3^{-1\frac{1}{2}} = 3^1 = 3$$

Exercise 1.11

In questions 1–6, evaluate the expressions. Do not use a calculator.

1 $5^{\frac{1}{2}} \times 5^{1\frac{1}{2}}$
2 $4^{\frac{1}{3}} \times 4^{\frac{1}{6}}$
3 $10^{\frac{1}{2}} \div 10^{1\frac{1}{2}}$
4 $49^{\frac{3}{4}} \div 49^{\frac{1}{4}}$
5 $4^{\frac{1}{3}} \times 2^{\frac{1}{3}}$
6 $16^{\frac{1}{3}} \div 4^{\frac{1}{6}}$

In questions 7–14, simplify the expressions.

7 $x^{\frac{1}{3}} \times x^{\frac{1}{6}}$

8 $\dfrac{a^{\frac{1}{2}} \times a^{\frac{1}{6}}}{a^{\frac{1}{3}}}$

9 $(x^{\frac{2}{3}})^6$

10 $\sqrt[3]{a^{1.5}}$

11 $\sqrt[3]{\dfrac{x^3}{y^6}}$

12 $(\sqrt[3]{a} \times \sqrt[6]{a})^2$

13 $\sqrt{b^{1\frac{1}{3}} \times b^{\frac{2}{3}}}$

14 $(\sqrt{x^{-5}})^{-4}$

SUMMARY

■ If x is not zero, then $x^0 = 1$. For example, $5^0 = 1$.

■ If x is not zero, then $x^{-n} = \dfrac{1}{x^n}$. For example, $2^{-3} = \frac{1}{8}$.

■ If x is not negative, then $x^{\frac{1}{n}} = \sqrt[n]{x}$, and $x^{\frac{m}{n}} = (\sqrt[n]{x})^m$. For example, $8^{\frac{1}{3}} = \sqrt[3]{8} = 2$.
■ For all values of the indices, powers obey the same rules for multiplication and division.

$$x^n \times x^m = x^{n+m} \qquad x^n \div x^m = x^{n-m} \qquad (x^n)^m = x^{nm} \qquad (xy)^m = x^n y^m \qquad \left(\dfrac{x}{y}\right)^n = \dfrac{x^n}{y^n}$$

Exercise 1A

1 Write $5^n \div 5^{-m}$ as a power of a single base.
2 Evaluate $(\frac{2}{3})^{-2}$.
3 Simplify $(2x)^4$.
4 Simplify the expression $121^n \div 11^m$.
5 Evaluate $81^{\frac{1}{4}}$ without the use of a calculator.
6 Use a calculator to evaluate $\sqrt[5]{11}$, giving your answer to 4 significant figures.
7 Evaluate $100^{\frac{3}{2}}$ without the use of a calculator.
8 Simplify the expression $x^{\frac{1}{8}} \times x^{\frac{7}{8}}$.
9 Simplify the expression $\sqrt[4]{x^2}$.
10 Evaluate the expression $121^{\frac{1}{4}} \times 11^{\frac{1}{2}}$ without a calculator.

Exercise 1B

1 Write $x^3 \times x^{-3}$ as a single power of x ($x \neq 0$).
2 What is the numerical value of your answer to question 1?
3 Simplify $(a^2 b^3)^4$.

4 Write $\dfrac{2^a}{16^{2b}}$ as a power of a single number.

5 Evaluate $64^{\frac{1}{3}}$ without the use of a calculator.
6 Use a calculator to evaluate $10^{\frac{2}{3}}$, giving your answer to 3 decimal places.
7 Simplify $\sqrt[5]{x^{10}}$.
8 Simplify $a^{\frac{3}{4}} \times a^{\frac{1}{4}}$.
9 Evaluate $36^{\frac{5}{8}} \div 6^{\frac{1}{4}}$ without the use of a calculator.
10 Write 1000 as a power of 100.

Exercise 1C Ma1

You already know about prime numbers. The first few prime numbers are 2, 3, 5, 7. This series never ends – there are infinitely many prime numbers, which means we can get prime numbers as large as we want.

Mathematicians like to find very large prime numbers. These numbers even have a practical use, in constructing secret codes. At the time of writing, the largest known prime is $2^{6972593} - 1$, discovered on 1 June 1999.

Try to enter this number on your calculator. What happens? If you wrote this number in full, how long would it be? Make a guess!

There is a rough conversion between powers of 2 and powers of 10.

$$2^{10} = 1024 \text{ and } 10^3 = 1000. \text{ So } 2^{10} \text{ is approximately equal to } 10^3.$$

What is the approximate value of $2^{6972593} - 1$, as a power of 10? How many digits does it have? How many pages would it take to write it out in full?

The record for the largest prime is broken frequently. Has $2^{6972593} - 1$ been beaten? You can use the internet to find the present record holder. Search for the key words *largest prime*.

> A kilobyte of computer memory is 1024 bytes. We say that a kilobyte is approximately 1000 bytes.

Exercise 1D Ma1

1 Which of the following are always true, and which can be false? For the true ones, find a justification. For the false ones, find a counterexample (that is, numerical values for which the rule doesn't hold).

a $(2ab)^2 = 2a^2b^2$ **b** $\sqrt{4x^4} = 2x^2$ **c** $\sqrt{8x^4} = 4x^2$

d $(a^4)^5 = a^9$ **e** $a^4 \times a^5 = a^9$ **f** $\sqrt[m]{(\sqrt[n]{x})} = \sqrt[n]{(\sqrt[m]{x})}$

2 What is wrong with the following 'proof' that $1 = -1$?

We know that $\dfrac{1}{-1} = \dfrac{-1}{1}$. Take the square root of both sides, then cross-multiply.

$$\frac{\sqrt{1}}{\sqrt{-1}} = \frac{\sqrt{-1}}{\sqrt{1}}$$

$$\sqrt{1} \times \sqrt{1} = \sqrt{-1} \times \sqrt{-1}$$

Hence $1 = -1$

You may be considering continuing with mathematics after GCSE. The next exercise is the first of several in which the material of the chapter is extended to topics that are studied after GCSE.

Exercise 1E

Logarithms were invented by a Scottish mathematician, John Napier (1550–1617) in 1594. He was also responsible for Napier's Bones, a device used for multiplying.
Use the internet to find out more about John Napier.

The inverse operation to taking powers is taking **logarithms**. As $2^3 = 8$, we write that 3 is the **logarithm of 8 to the base 2**. It is written as $3 = \log_2 8$. In general, if $a^n = x$, then we write $n = \log_a x$.

1 Find these.

   a $\log_{10} 100$ **b** $\log_3 9$ **c** $\log_{10} 1000$

   d $\log_4 2$ **e** $\log_{10} 0.1$ **f** $\log_{100} 0.1$

2 There are rules for multiplying and dividing powers. Can you find the corresponding rules for logarithms? Try to find rules for:

   a $\log_a xy$ **b** $\log_a \dfrac{x}{y}$ **c** $\log_a x^n$

3 Before the days of calculators, a lot of calculation was done using logarithms. Can you see from your answer to question 2 how it was done?

2 Probability

Events and outcomes

The word **event** is used in sport. Football matches or horse races are events. In mathematics we use the word more generally. Any experiment that has at least one possible result is called an event. The following are all events:

- A 100 m race
- A netball match
- The rolling of a dice

The **outcomes** of the events are the different results. For the 100 m race, the outcomes are the possible winners. For the netball match, the outcomes are a win, a loss and a draw. For the dice, the outcomes are the numbers 1–6.

For these events, the result is not certain. The probability of an outcome is a measure of how likely we think it is. Probability runs from 0 (impossible) to 1 (certain). If an outcome may or may not happen, then its probability lies between 0 and 1.

If all the outcomes of an event are equally likely, then you can work out their probabilities. If a fair dice is rolled, the probability of it giving us a 6 is $\frac{1}{6}$. Sometimes you can estimate probability from experience; you might know, for example, that the probability that a day in August will be dry is $\frac{3}{4}$. But, for the 100 m race and the netball match, the probabilities that you give for the outcomes depends on your own opinions.

These are all single outcomes. Outcomes can be combined, with the words *not*, *or* and *and*.

Not

If the probability of an outcome is p, then the probability of the outcome not happening is $1 - p$.

$$P(\text{not } A) = 1 - P(A)$$

So, for example, if the probability of Angela winning the 100 m race is $\frac{1}{8}$, then the probability that she won't win the race is $\frac{7}{8}$.

Or

Two outcomes are **mutually exclusive** if they cannot both happen together. The probability of either of two mutually exclusive outcomes happening is the sum of the individual probabilities.

$$P(A \text{ or } B) = P(A) + P(B) \quad \text{provided } A \text{ and } B \text{ are mutually exclusive}$$

If, in a netball match, the probability of a win is $\frac{1}{3}$ and the probability of a loss is $\frac{5}{12}$, then the probability of either a win *or* a loss is $\frac{1}{3} + \frac{5}{12}$, which is $\frac{3}{4}$.

Note. Before adding probabilities, be sure to check that the outcomes are mutually exclusive, that is, that they cannot happen together. Suppose the event is drawing a card from a pack. Consider these outcomes.

A The card is a Spade
B The card is a Diamond
C The card is a Queen.

Outcomes *A* and *B* are mutually exclusive, because the card cannot be both a Spade and a Diamond. But events *A* and *C* are not mutually exclusive, because the card could be the Queen of Spades.

The Venn diagram shows what can happen. Notice that the Spade and Diamond regions are separate, but the Spade and Queen regions overlap.

Venn diagrams were named after an English mathematician called John Venn (1834–1923).

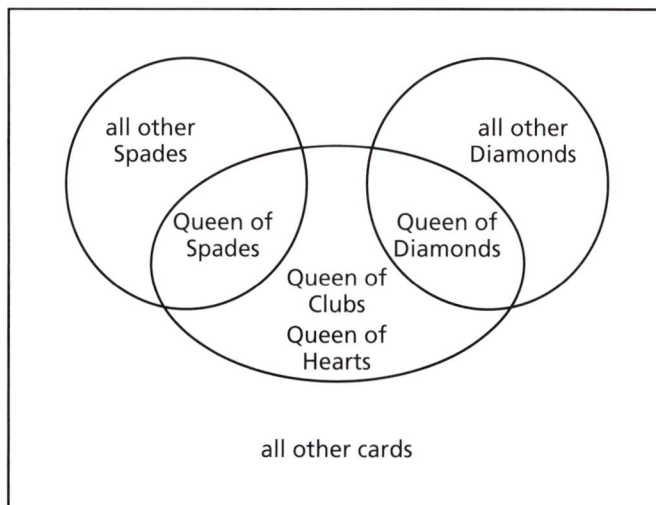

Example   Damien reckons that his probability of getting grade A at Maths GCSE is 0.4, and that his probability of getting grade B is 0.3. What does he reckon will be his probabilities of:

a not getting grade A?   **b** getting either grade A or grade B?

The probability of not getting grade A is 1 minus the probability of getting grade A.

$$1 - 0.4 = 0.6$$

So Damien reckons the probability of not getting grade A is 0.6.
   Damien cannot be awarded different grades for the same exam. So the outcomes are mutually exclusive. To find the probability of getting grade A or grade B, add the probabilities.

$$0.4 + 0.3 = 0.7$$

The probability of getting either grade A or grade B is 0.7.

Exercise 2.1

1 If the probability that it will be dry tomorrow is $\frac{3}{4}$, what is the probability that it won't be dry?

2 A coin is misshapen, so that the probability of getting Heads is 0.4. What is the probability of getting Tails?

> A coin like this is called *biased.*

3 The probability that Amy will pass her driving test is $\frac{5}{8}$. What is the probability that she won't pass it?

4 Consider the event of drawing a card from a pack. Some outcomes are listed below. Which pairs of outcomes are mutually exclusive?
   A The card is a Heart.
   B The card is a Club.
   C The card is black.
   D The card is an Ace.

5 Consider the event of rolling a dice. Some outcomes are listed below. Which pairs of outcomes are mutually exclusive?
   A The result is 5.
   B The result is 4.
   C The result is even.
   D The result is prime.

6 Consider the event of rolling two dice. Some outcomes are listed below. Which pairs of outcomes are mutually exclusive?
   A The total is 7.
   B The total is 12.
   C There is a 'double' (both dice with the same number).
   D The first dice is 3.
   E The second dice is 6.

7 Ballytrim and Whitefoot are in the same amateur soccer league. Their chances of winning the league are $\frac{1}{10}$ and $\frac{1}{12}$ respectively. What is the probability that one of the teams will win?

8 In the situation of question 7, what is the probability that neither Ballytrim nor Whitefoot will win their league?

9 Sean and Darren have probabilities $\frac{1}{3}$ and $\frac{1}{6}$ respectively of being picked as the opening batsman for their school's cricket team. What is the probability that one of them will be chosen?

10 In the situation of question 9, what is the probability that neither Sean nor Darren will be picked as the opening batsman?
11 Siobhan has 0.8 chance of passing English GCSE, and 0.6 chance of passing Maths GCSE. Can you find the probability that she will pass either English or Maths?
12 For an athletics competition, Ethan has entered the 100 m and 200 m races. His chances of winning the races are $\frac{1}{10}$ and $\frac{1}{8}$ respectively. Can you find the probability that he will win either of the races?

Multiplying probabilities

A fair coin is spun and a fair dice is rolled. The coin can be Heads or Tails (H or T) and the dice can be any number from 1 to 6. We can list the outcomes as follows:

H1 H2 H3 H4 H5 H6
T1 T2 T3 T4 T5 T6

Both coin and dice are fair. So each of these 12 outcomes is equally likely; each has probability $\frac{1}{12}$. Note also that the coin and the dice do not affect each other. The result of the spinning of the coin does not affect the result of the rolling of the dice.

Look at the list of outcomes above. There are two rows and six columns. Hence the probability of getting T is $\frac{1}{2}$, and the probability of getting 4 is $\frac{1}{6}$. The product of these is $\frac{1}{2} \times \frac{1}{6}$, which is $\frac{1}{12}$. The probability of both outcomes happening is the product of the individual probabilities.

In general, two outcomes A and B which do not make each other either more or less likely (i.e. they are not connected) are called **independent**. If A and B are independent, the probability of them both occurring is the product of the individual probabilities.

$$P(A \text{ and } B) = P(A) \times P(B) \quad \text{provided } A \text{ and } B \text{ are independent}$$

As a justification, look at the example above. When we spin the coin, on half the spins we expect to get a Tail. When we roll the dice, on a sixth of the rolls we expect to get a 4. So when we do both, we expect to get a Tail and a 4 on a sixth of a half of the trials, that is, on $\frac{1}{6} \times \frac{1}{2}$ of the trials.

Exercise 2.2

1 Two fair spinners each have three edges. The edges of the first spinner are labelled A, B and C, and the edges of the second spinner are labelled 1, 2 and 3. When both are spun, one possible outcome is A1 (A on the first spinner, 1 on the second).
   a List all the possible outcomes.
   b What is the probability of outcome C2?
   c What are the probabilities of C on the first spinner, and of 2 on the second spinner? Is the product of these probabilities equal to your answer to **b**?
2 A fair tetrahedral (four-sided) dice has the numbers 1–4 on its faces. The dice is rolled and a fair coin is spun. If the dice lands on the face labelled 1, and the coin lands Heads up, then the outcome is recorded as 1H.
   a List all the possible outcomes.
   b What is the probability of outcome 3T?
   c What are the probabilities of 3 on the dice and of T on the coin? Is the product of these probabilities equal to your answer to **b**?

3 The probability that the racehorse Zohar will win its race this afternoon is $\frac{1}{8}$, and the probability that Barchester City will win its football match is $\frac{2}{9}$. What is the probability that both will win?

4 In the situation of question 3, find the probabilities that:
   a Zohar does not win the race
   b Barchester City does not win the match
   c Zohar wins and Barchester City loses
   d Zohar loses and Barchester City wins
   e both Zohar and Barchester City lose.
   Add up your answers to parts **c**, **d** and **e**. Add this to your answer to question 3. Is the result 1?

5 Jane has a French penpal Alliette. The probability that Jane will pass her Maths GCSE is 0.7, and the probability that Alliette will pass the Maths component of the *'Bac'* is 0.6. What is the probability that they will both pass?

6 In the situation of question 5, find the probabilities that:
   a Jane does not pass
   b Alliette does not pass
   c Jane passes but Alliette does not
   d Jane does not pass but Alliette does
   e neither passes.
   Add up your answers to parts **c**, **d** and **e**. Add this to your answer to question 5. Is the result 1?

Tree diagrams

Amena goes to a theme park. A map of part of the park is shown below.

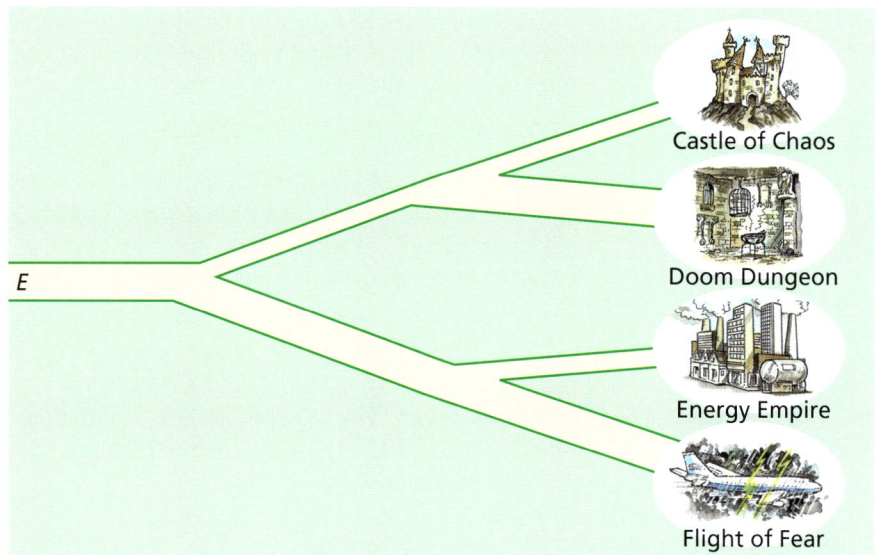

The entrance is at *E*. It has been noticed that at the first junction visitors tend to turn left with probability $\frac{1}{4}$, and at either of the second junctions they tend to turn left with probability $\frac{1}{3}$. Where will Amena end up?

She has four possible journeys:

Left then left, ending up at the Castle of Chaos
Left then right, ending up at the Doom Dungeon
Right then left, ending up at the Energy Empire
Right then right, ending up at the Flight of Fear.

The probabilities of each journey are found by multiplying the probabilities at each junction. For example:

$$P(\text{end at the Doom Dungeon}) = P(\text{left at first junction}) \times$$
$$P(\text{right at second junction})$$
$$= \tfrac{1}{4} \times \tfrac{2}{3}$$
$$= \tfrac{2}{12}$$
$$= \tfrac{1}{6}$$

Exercise 2.3

1 Find the probability that Amena from the example above will end up at:
   a the Castle of Chaos   **b** the Energy Empire   **c** the Flight of Fear.
2 Add the four probabilities found so far. Do they come to 1?

We can simplify the map of the park, and on each path write the probability that Amena will take that path. It will look like this.

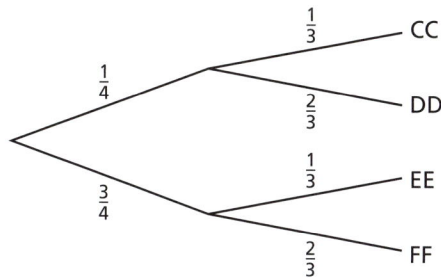

Remember:

The probabilities are multiplied because Amena will go left and then right.

It is called a tree diagram, but as it is horizontal it's more like a tree blown over by the wind!

To find the probability that Amena will take a particular route, multiply the probabilities along the route. For the route that ends up at the Doom Dungeon, the probability is $\tfrac{1}{4} \times \tfrac{2}{3}$, which is $\tfrac{1}{6}$.

   This diagram is called a **tree diagram**, which is often very useful to show all the possible results of two or more experiments.

   The first fork of the tree diagram shows the results of the first experiment. Put the probability of each result alongside its branch. The second fork shows the results of the second experiment. Again, put the probability of each result alongside its branch.

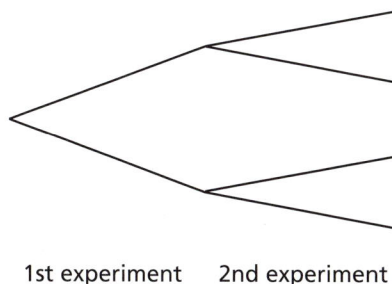

1st experiment   2nd experiment

To find the probability of two results, multiply the probabilities along the relevant branches. Write the result at the end of the branch.

Examples   The tree diagram shows what happens when two misshapen coins are spun. What is the probability that two Heads are obtained?

1st coin   2nd coin

Two Heads corresponds to the top branch. Multiply the probabilities along this branch.

$$\tfrac{1}{3} \times \tfrac{1}{4} = \tfrac{1}{12}$$

The probability of two Heads is $\tfrac{1}{12}$.

Joe has two boxes: the first contains 3 white counters and 4 black counters, and the second contains 2 white counters and 7 black counters. He draws one counter from each box. Draw a tree diagram, and find the probability that he draws two white counters.

The first experiment is drawing a counter from the first box. There are two possible results, either a white or a black counter is drawn, with probabilities $\tfrac{3}{7}$ and $\tfrac{4}{7}$ respectively. Put these probabilities on the lines leading from the first fork.

The second experiment is drawing a counter from the second box. Again there are two possible results, either a white or a black counter is drawn, with probabilities $\tfrac{2}{9}$ and $\tfrac{7}{9}$ respectively. Put these probabilities on *both* the lines leading from the second fork.

At the ends of the branches write the outcomes, WW, WB, BW and BB.

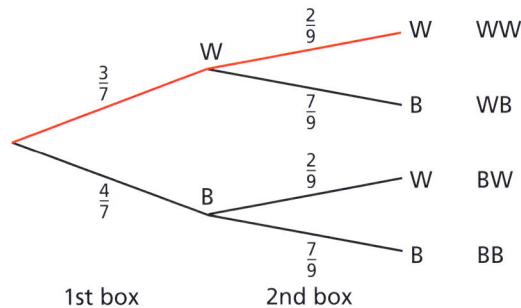

1st box   2nd box

The branch corresponding to two white counters is the top branch, labelled WW, and shown in red. Multiply the probabilities along this branch.

$$\begin{aligned}
\text{P(two white counters)} &= \tfrac{3}{7} \times \tfrac{2}{9} \\
&= \tfrac{6}{63} \\
&= \tfrac{2}{21}
\end{aligned}$$

The probability of getting two white counters is $\tfrac{2}{21}$.

Exercise 2.4

1 In the first example on page 18 about misshapen coins, we found that the probability that both coins were Heads was $\frac{1}{12}$. Now use the diagram to:
  a Find the probability that both coins are Tails.
  b Find the probability that the first coin is Heads and the second is Tails.
  c Find the probability that the first coin is Tails and the second is Heads.
  d Add the four probabilities found so far. Do they come to 1?

2 In the second example on page 18 about drawing counters, we found that the probability of two white counters was $\frac{2}{21}$.
  a Find the probability of drawing a white counter then a black counter.
  b Find the probability of drawing a black counter then a white counter.
  c Find the probability of drawing two black counters.
  d Add the four probabilities found so far. Do they come to 1?

3 Two incomplete packs of cards have 45 and 50 cards respectively. One card is drawn from each. The tree diagram shows whether the cards were red or black.

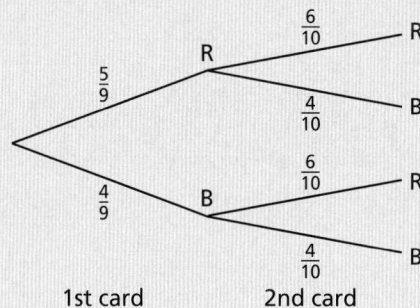

  a Find the probability that both cards were black.
  b Find the probability that both cards were red.
  c Find the probability that the first card was red and the second black.
  d Find the probability that the first card was black and the second red.
  e Do your answers add up to 1?

4 Jack is watching the people who make calls from a phone booth. He finds that $\frac{1}{3}$ of the callers are men. Copy and complete the tree diagram below for two successive callers.

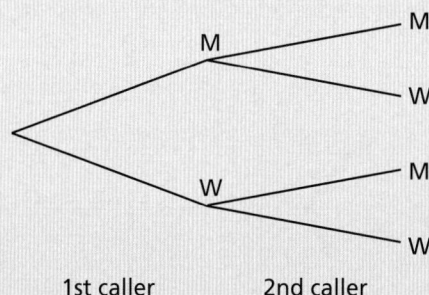

  a What is the probability that both callers are women?
  b What is the probability that both callers are men?
  c What is the probability that the first caller is a man and the second a woman?
  d What is the probability that the first caller is a woman and the second a man?
  e Do your answers add up to 1?

Sometimes we add the probabilities of branches. Recall from page 13 that two events are mutually exclusive if they cannot happen together. Clearly the branches of a tree diagram are mutually exclusive. Amena cannot end up at both the Doom Dungeon and the Castle of Chaos! So the probability that she ends up at one or the other is the sum of the individual probabilities.

Example

Look again at the example on page 18 about drawing counters from boxes. Suppose that Joe is interested in getting just one white counter. What is the probability of this? There are two ways he could get just one white counter.

$$\frac{3}{7} \times \frac{7}{9} = \frac{1}{3}$$
$$\frac{4}{7} \times \frac{2}{9} = \frac{8}{63}$$

WB: white counter then black counter. This has probability $\frac{1}{3}$.
BW: black counter then white counter. This has probability $\frac{8}{63}$.

These events cannot happen together. So the probability of either happening is the sum of the probabilities.

$$\frac{1}{3} + \frac{8}{63} = \frac{21}{63} + \frac{8}{63}$$

P(white then black or black then white) $= \frac{1}{3} + \frac{8}{63}$
$$= \frac{29}{63}$$

The probability of just one white counter is $\frac{29}{63}$.

Exercise 2.5

1 A biased coin is spun twice. The tree diagram shown gives the probabilities of the different outcomes.

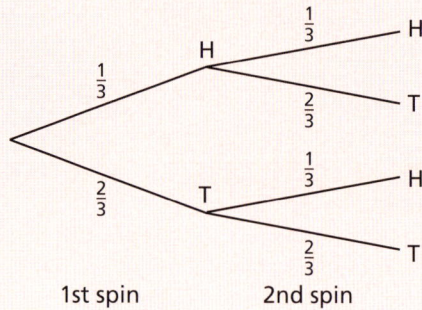

1st spin   2nd spin

What is the probability that
a two Heads were obtained   **b** exactly one Head was obtained?

2 A biased dice is rolled twice. The tree diagram below shows the probabilities of obtaining sixes.

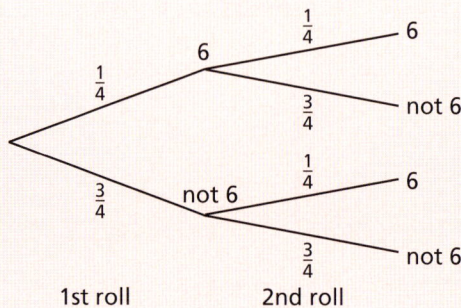

1st roll   2nd roll

What is the probability that
a two sixes were obtained   **b** exactly one six was obtained?

3 Nadine has two packs of cards. The first is a standard pack with 52 cards, but the second is incomplete, with only 9 Hearts out of 48 cards. She draws one card from each pack. Copy and complete the tree diagram shown.

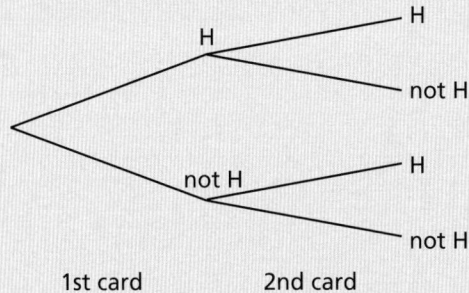

1st card   2nd card

What is the probability that
a she gets two Hearts   **b** she gets no Hearts   **c** she gets exactly one Heart?

4 There are two classes in Year 9: the first class has 10 boys and 12 girls, and the second class has 13 boys and 11 girls. One pupil is picked from each class. Copy and complete the tree diagram shown.

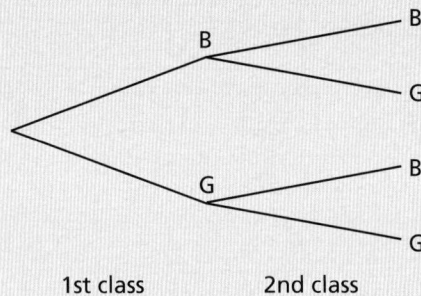

1st class   2nd class

What is the probability that exactly one girl is picked?

5 There are two fish tanks in a restaurant. The first contains 8 fish, of which 3 are trout, and the second contains 10 fish, of which 7 are trout. One fish is picked at random from each tank. Copy and complete the tree diagram shown.

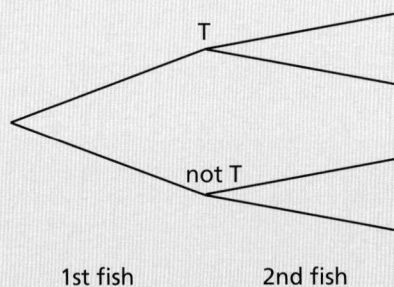

1st fish   2nd fish

What is the probability that exactly one trout is picked?

6 An octahedral (eight-sided) dice has the numbers 1–8 on its faces. It is rolled, and afterwards an ordinary six-sided dice is rolled. Copy and complete the tree diagram shown.

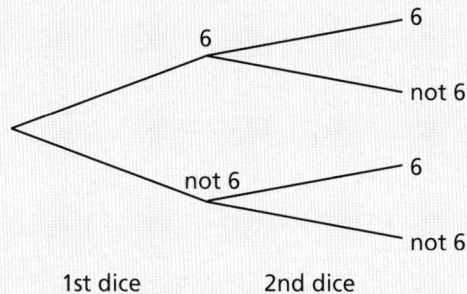

1st dice   2nd dice

Find the probability of
a one six   **b** no sixes   **c** at least one six.

7 When Jenny and Aziz take their driving test, they have probabilities 0.7 and 0.6 respectively of passing it. Draw a tree diagram to show the situation, and hence find the probability that
a both pass   **b** both fail   **c** exactly one passes.

8 Two footballers each take a penalty shot at goal. Their chances of scoring are 0.8 and 0.7 respectively. Draw a tree diagram to show the situation, and hence find the probability that
a both score   **b** neither scores   **c** exactly one scores.

9 When a tennis player serves from the right-hand court, her probability of a good first serve is 0.5. When serving from the left-hand court, her probability is 0.45. Draw a tree diagram to illustrate her first two serves, and hence find the probability that
a both first serves are good   **b** exactly one first serve is good.

10 When Damien takes a driving test, he has probability $\frac{1}{3}$ of passing. If he fails he takes the test again, but he doesn't get any better or worse at subsequent attempts. Copy and complete the tree diagram below.

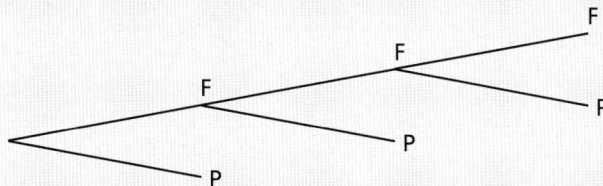

What is the probability that Damien has
a one driving test   **b** two driving tests   **c** at least three tests
before he passes?

Repeated trials

If you spin a fair coin twice, the probability of two heads is $\frac{1}{2}\times\frac{1}{2}$, which is $\frac{1}{4}$. If you spin it three times, the probability of three heads is $\frac{1}{2}\times\frac{1}{2}\times\frac{1}{2}$, which is $(\frac{1}{2})^3$, or $\frac{1}{8}$. You don't have to draw a tree diagram for this. In general, if you spin a coin n times, the probability of n heads is $(\frac{1}{2})^n$.

In 10 spins, the probability that all 10 are heads is $(\frac{1}{2})^{10}$, which is less than 0.001. So the probability of *not* getting 10 heads is $1-(\frac{1}{2})^{10}$, which is over 0.999. The probability of at least one tail in the 10 spins is over 0.999.

Example When Lydia takes her driving test, she has probability $\frac{3}{8}$ of passing. This probability does not change for subsequent attempts. What is the probability that she will pass within six attempts?

The probability that she will *fail* the test is $1-\frac{3}{8}$, which is $\frac{5}{8}$. She will pass within six attempts provided she does not fail all of them.

$$P(\text{fail all six}) = (\tfrac{5}{8})^6$$
$$P(\text{pass within six attempts}) = 1 - P(\text{fail all six attempts})$$
$$= 1 - (\tfrac{5}{8})^6$$
$$= 0.94$$

The probability that she will pass within six attempts is 0.94.

Exercise 2.6

1 A fair coin is spun eight times. What is the probability of eight Heads?
2 A fair coin is spun seven times. What is the probability of at least one Head?
3 A fair dice is rolled five times. What is the probability of no sixes?
4 A fair dice is rolled ten times. What is the probability of at least one six?
5 A three-edged spinner has the letters A, B and C on its edges.
   a If it is spun seven times, what is the probability of seven As?
   b If it is spun eight times, what is the probability of no As?
   c If it is spun nine times, what is the probability of at least one A?
   d If it is spun six times, what is the probability that at least one spin is not an A?
6 A tetrahedral (four-sided) dice has the numbers 1–4 on its faces. It is rolled six times. Find the probabilities of the following outcomes.
   a It lands on the face numbered 4 every time.
   b It never lands on the face numbered 4.
   c It lands on the 4 face at least once.
   d At least once it lands on a face other than the 4 face.
7 The chance of winning a prize in the National Lottery is about $\frac{1}{50}$. If you enter every Saturday for a year, what is the probability that you will win at least one prize?
8 When Ben throws a dart at a dartboard, the probability that it will land in the bullseye is $\frac{1}{10}$. What is the probability of at least one bullseye in eight throws?
9 When Val serves at tennis, the chance her first serve will be good is 0.55. If she serves eight times, find the probability that at least one first serve will be good.

Finding the number of trials

In some games you have to roll a dice at the beginning, and you cannot start until you have rolled a 6. How many rolls will it take?

In theory, there is no limit to the number of rolls you need. But you can be almost certain that you will get a 6 within a large number of tries. For example, you are 99% certain of getting a 6 within 26 tries.

Example A fair coin is spun repeatedly. How many spins are needed for us to be 90% sure of a Head?

The chance of *not* getting a Head is $\frac{1}{2}$. Suppose the coin is spun n times. The probability that we don't get a Head in any of the spins is $(\frac{1}{2})^n$. We want this to be smaller than 0.1.

$$\frac{1}{2^n} < 0.1$$

Try values of n until $\frac{1}{2^n}$ is less than 0.1.

$$\frac{1}{2^2} = 0.25 \quad \frac{1}{2^3} = 0.125 \quad \frac{1}{2^4} = 0.0625 < 0.1$$

Four spins are needed.

Exercise 2.7

1 Marvin has probability $\frac{2}{3}$ of passing his driving test. How many tests must he take to be 90% sure of passing the test?
2 The probability that a tulip bulb will flower is $\frac{8}{9}$. How many bulbs must I buy to be 99% sure that at least one will flower?
3 A quarter of the silicon chips produced by an electronics firm are faulty. How many chips must it produce to have probability 0.99 that at least one is non-faulty?
4 Above it was stated that if you roll a fair dice 26 times you are 99% sure of at least one 6. Verify this result.
5 Verify that if you spin a fair coin seven times you are 99% sure of at least one Head.
6 One in ten bottles of Swilch contains a prize token. Verify that if you buy 22 bottles you are 90% sure of getting at least one prize.
7 The probability that an entry in the National Lottery will win a prize is about $\frac{1}{50}$. Verify that if you enter 35 times you have a better than 0.5 probability of winning at least one prize.

Exercise 2.8 (Ma1)

You will need:
- a coin

In this section we found how many repeated trials were needed to achieve a certain result. This exercise involves finding the probabilities by experiment. Work with a partner.

Spin a coin until a Head is obtained. Record how many spins were needed. For example, if the result was TTH, then three spins were needed.

Repeat this many times. Record the results in a table like the one below.

number of spins to get a Head	1	2	3	4	5	6	7	8
frequency								

From the table, find the experimental probability that a certain number of spins are required. Does it agree with the theoretical probability?

SUMMARY

- If two **outcomes** of an **event** cannot both happen (they are **mutually exclusive**), then the probability that either of them happens is the sum of their individual probabilities. $P(A \text{ or } B) = P(A) + P(B)$.
- If two outcomes do not affect each other (they are **independent** events), then the probability of both happening is the product of their individual probabilities. $P(A \text{ and } B) = P(A) \times P(B)$.
- Combinations of probabilities can be shown on a **tree diagram**.
- Suppose a trial is repeated many times. The probability of at least one success is 1 minus the probability of a string of failures.
- By taking powers, you can find how many trials are needed to be almost certain of getting a success.

Exercise 2A

1 An icosahedral (20-faced) dice has the numbers 1–20 on its faces. It is rolled once.
   a What is the probability that it doesn't land on the face labelled 17?
   b What is the probability that it lands on a face labelled 6 or 12?
2 A dodecahedral (12-faced) dice has the numbers 1–12 on its faces. It is rolled at the same time as the icosahedral dice of question 1. Find the probability that:
   a both dice land on the face labelled 10
   b neither dice lands on the face labelled 10
   c at least one of the dice lands on the face labelled 10.

3 In a multiple choice test, each question has four possible answers. David answers all the questions at random.

a Copy and complete this tree diagram for the first two questions.

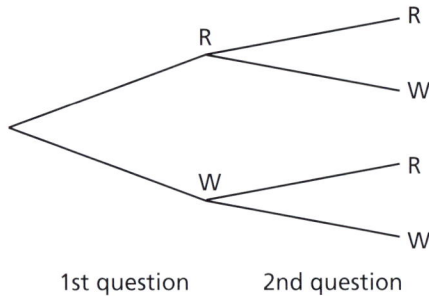

1st question   2nd question

b What is the probability that he gets both questions right?
c What is the probability that he gets exactly one question right?

4 In the situation of question 3, what is the probability that David gets at least one of the first ten questions right?

5 In the situation of question 3, how many questions must David answer before he is 99% sure of getting at least one wrong?

Exercise 2B

1 A roulette wheel has the holes numbered 1–36. It is spun once.
a What is the probability that the ball doesn't land in the hole numbered 19?
b What is the probability that the ball lands in a hole numbered 4 or 34?

2 The roulette wheel of question 1 is spun twice. For the first spin, I bet that an even number will come up. For the second spin, I bet that a number between 1 and 12 inclusive will come up.
a What is the probability that I win both bets?
b What is the probability that I win neither bet?
c What is the probability that I win at least one bet?

3 The lunch menu of a restaurant contains four choices for the first course, of which one is vegetarian. There are seven choices for the second course, of which two are vegetarian. A diner picks a first and second course at random.
a Copy and complete this tree diagram.

1st course   2nd course

b What is the probability that both courses are vegetarian?
c What is the probability of at least one vegetarian course?

4 When Tina plays Ishtar at tennis, the chance that Tina wins a point is $\frac{5}{8}$. What is the probability that in five points Ishtar will win at least one?

5 In the situation of question 4, how many points must be played before Tina is 95% sure of winning at least one?

Exercise 2C Ma1

The study of probability took off in the seventeenth century, after an exchange of letters between two French mathematicians, Pierre Fermat and Blaise Pascal. In particular, they discussed the following gambling problem, called the *Problem of the points*.

> Two players of a dice game have equal skill. They play alternate points, and the first to gain 6 points wins. The game is interrupted when A has 4 points and B has 3. How should the stakes be divided?

1 Suppose the game could be continued. Show that the game would end after at most four more points.

2 One possible result of the next four points could be written $ABBA$. Then A would win, having reached 6 points. List all the other possible results.

3 In how many of the results would A win, and in how many would B win? In what ratio would you divide the stakes?

4 Use this method to divide the stakes if A has won 1 point and B has won 5.

This was Fermat's solution to the problem. Pascal's solution involved the triangle of numbers known as Pascal's triangle. Use the internet to find out about Pascal and Fermat.

Exercise 2D Ma1

If outcomes A and B are independent, then $P(A \text{ and } B) = P(A) \times P(B)$. Going the other way, if $P(A \text{ and } B) \neq P(A) \times P(B)$, then A and B are not independent. This can be very important in medical research, for example, suppose A represents eating a particular food, and B represents contracting a particular medical condition. If $P(A \text{ and } B) > P(A) \times P(B)$, then A and B are not independent, and there is a connection between eating the food and contracting the condition.

Investigate for yourself whether two qualities A and B are connected. Collect data, and from it find the experimental probabilities of having quality A, having quality B and having *both* qualities. Is it true that $P(A \text{ and } B) = P(A) \times P(B)$? You could try one of these.

> Are left-handed people better at a particular sport?
> Do short-sighted people read more books?

Exercise 2E

In the final section of this chapter you found how many trials were needed to obtain a certain result. If only a few trials are needed, then you can find the number by taking successive powers. If the number is large, this may take a long time. How could we answer the question: 'How many rolls of a dice are needed before we are 99% sure that at least one will give a 6?'

1 A computer can do all the work of taking successive powers. In particular, a spreadsheet can be set up to answer the question above. To find how many rolls of a dice are needed to be 99% sure of a 6, we need to find n so that $(5/6)^n < 0.01$.

Enter 5/6 in cell A1. You may find that the computer thinks that this is a date, that is, 5 June, in which case enter =(5/6). In cell A2 enter =(5/6)*A1. The entries should be like this:

Copy the formula in A2 down the A column as far as you like. In the nth row you have $(5/6)^n$. For what value of n is this less than 0.01?

2   Adapt your spreadsheet to find how many spins of a coin are needed to be 99% sure of at least one Head.

3   In a multiple choice test, each question has five possible answers. If a candidate answers at random, how many questions must he try before he is 99% sure of getting at least one right?

The next question extends the material of this chapter to a topic beyond GCSE.

4   Logarithms were defined in exercise 1E. On your calculator you will find a button labelled log. It gives logarithms to the base 10. One law of logarithms is

$$\log_{10} a^n = n \times \log_{10} a$$

Take logarithms of both sides of the inequality $(5/6)^n < 0.01$. Use the law above to find n. Use this method to solve questions 2 and 3, without the help of a computer.

3 Area and volume

Volumes of prisms

A **prism** is a solid whose cross-section is constant. The shape we usually refer to as a prism has a triangular cross-section, for example prism binoculars use these prisms to turn the image the right way up. But the definition applies to any shape with constant cross-section. Some pencils, when unsharpened, are prisms with a hexagonal cross-section.

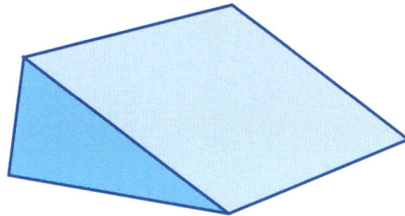

There are several common shapes which are prisms, even if we do not think of them as such. A cuboid has a rectangular cross-section, and a cylinder has a circular cross-section. So both of these solids are prisms.

For any prism, the volume is found by multiplying the area of cross-section by the height.

$$V = A \times h$$

Examples   Find the volume of a cylinder with height 10 cm and base radius 3 cm.

The cross-section is a circle of radius 3 cm. The area of cross-section is πr^2, that is, $\pi \times 3^2 \, \text{cm}^2$.

$$\text{volume} = 10 \times \pi \times 3^2 \, \text{cm}^3$$
$$= 283 \, \text{cm}^3$$

The volume is 283 cm³.

....

A cuboid has a square base and height 8 cm. If its volume is 120 cm³, find the side of the base.

Let the side be x cm. Then

$$x^2 \times 8 = 120$$
$$x^2 \qquad = 15$$
$$x \qquad = \sqrt{15}$$

The side of the base is 3.87 cm.

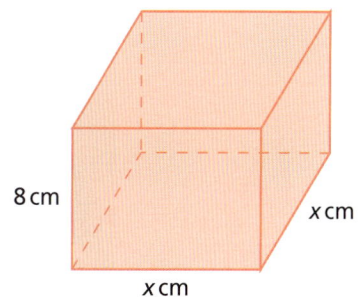

8 cm

x cm

x cm

Exercise 3.1

1 Find the volumes of these solids.

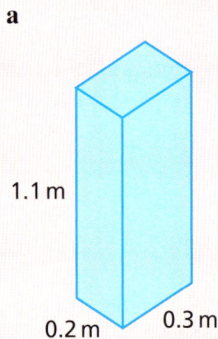

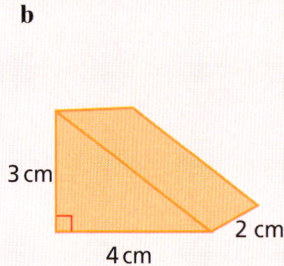

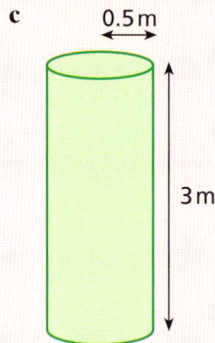

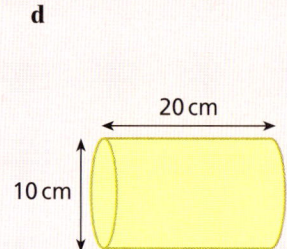

a

1.1 m

0.2 m   0.3 m

b

3 cm

4 cm   2 cm

c   0.5 m

3 m

d

20 cm

10 cm

2 A prism is 7 cm long, and has a cross-section which is a right-angled triangle with sides 3 cm, 4 cm and 5 cm. Find
   a the volume of the prism   **b** the surface area of the prism.

3 The cross-section of a prism is a right-angled triangle with sides 8 cm, 15 cm and 17 cm. The prism is 20 cm long. Find
   a the volume of the prism   **b** the surface area of the prism.

4 Find the volume of a cylinder with base radius 0.2 m and height 0.4 m.

5 Find the volume of a cylinder with base radius 22 cm and height 41 cm.

6 A cylinder has base radius 5 cm and volume 160 cm^3. Find its height.

7 Copper wire has a cross-section which is a circle with radius 0.5 mm. What length of wire can be made from 400 cm^3 of copper?

8 A cuboid has a square base, its height is 0.8 m and its volume is 1.4 m^3. Find the side of the base.

9 A cylinder has height 1.3 m and volume 5 m^3. Find its base radius.

10 100 cm^3 of copper is made into wire with circular cross-section. If the wire is 200 m long, find the radius of cross-section.

Volume of a pyramid and a cone

Not all shapes have constant cross-section. There are examples of solids which slope from their base to a point. There are solids which slope uniformly, that is, the lines from the base to the point are straight.

A **pyramid** slopes uniformly from a base which is a polygon. The famous pyramids of Egypt have square bases.

A **cone** slopes uniformly from a circle to a point.

For all these solids, the formula for their volume is

$$\text{volume} = \tfrac{1}{3}\,\text{base area} \times \text{height}$$
$$V = \tfrac{1}{3}Ah$$

So, if a cone has height h and base radius r, its base area is πr^2. Its volume V is given by

$$V = \tfrac{1}{3}\pi r^2 h$$

A complete proof of this result is beyond the scope of this book. Here is a justification.

Justification

Consider a cube $ABCDEFGH$ of side $2a$. Let X be the point in the middle of the cube. The cube can be cut into six equal pyramids, formed by joining X with each of the six faces. For example, one of the pyramids is $XABCD$.

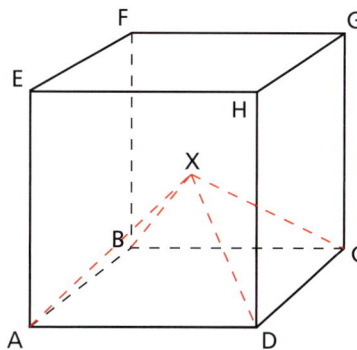

$$\text{volume of } XABCD = \tfrac{1}{6} \times \text{volume of cube} = \tfrac{1}{6} \times (2a)^3 = \tfrac{8}{3}a^3$$

The base of $XABCD$ is a square of side $2a$, and its height is a. Using the formula above for the volume of a pyramid

$$\tfrac{1}{3}Ah = \tfrac{1}{3} \times (2a)^2 \times a = \tfrac{8}{3}a^3$$

The two results are the same.

This verifies the formula. It does not *prove* it, as we have only shown it to be true for one particular pyramid.

Examples Find the volume of a cone of height 6 cm and base radius 2 cm.

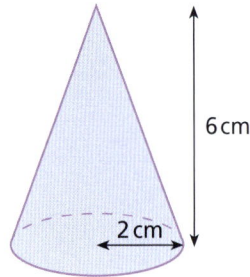

Using the formula

$$V = \tfrac{1}{3}\pi r^2 h = \tfrac{1}{3} \times \pi \times 2^2 \times 6 = 25.1$$

The volume is 25.1 cm³.

. . . .

A square-based pyramid has height 10 cm and volume 48 cm³. Find the side of the base.

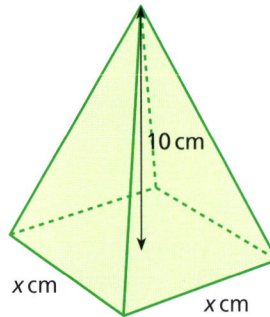

Suppose the side is *x* cm. Use the formula

$$\tfrac{1}{3}x^2 10 = 48$$
$$10x^2 = 144$$
$$x^2 = 14.4$$

multiplying by 3 to clear
up the fraction

Hence $x = \sqrt{14.4}$.

The side of the base is 3.79 cm.

Exercise 3.2

1 Find the volume of a pyramid whose base is a square of side 6 cm and of height 10 cm.
2 Find the volume of a pyramid whose base is a rectangle 3 m by 4 m, and whose height is 2.5 m.
3 A pyramid has a rectangular base which is 5 m by 6 m. If the volume is 200 m³, find the height.
4 A pyramid has a square base of side 18 cm. If its volume is 1000 cm³, find its height.
5 A pyramid with a square base has height 7 m and volume 84 m³. Find the side of its base.
6 A pyramid with a square base has height 5.5 cm and volume 200 cm³. Find the side of its base.
7 Find the volume of a cone with base radius 3 cm and height 8 cm.
8 A cone has base radius 0.7 m and height 1.3 m. Find its volume.
9 A cone has base radius 5 cm and volume 27 cm³. Find its height.
10 Find the height of a cone which has volume 1.3 m³ and base radius 0.6 m.
11 A cone has height 30 cm and volume 40 cm³. Find its base radius.
12 Find the base radius of a cone with height 4 m and volume 20 m³.

Volume and surface area of a sphere

If a sphere has radius r, then its volume is given by

$$V = \tfrac{4}{3}\pi r^3$$

The surface area of the sphere is given by

$$A = 4\pi r^2$$

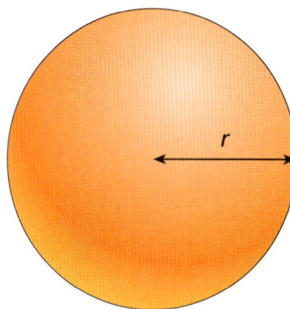

These results were discovered by the Greek mathematician Archimedes. Exercise 3E at the end of this chapter gives a simplified version of the proof for the formula for the volume.

> Archimedes lived from 287 to 212 BC. He was killed when the Romans seized Syracuse – he was ordered to go and see the Roman commander, but refused because he was in the middle of working out a problem!

Examples

Find the volume of a sphere of radius 4 m.

Let the volume be V m^3. Using the formula

$$V = \tfrac{4}{3} \times \pi \times 4^3$$
$$= 268$$

The volume is 268 m^3.

The volume of a sphere is 20 cm^3. Find its surface area.

First we must find the radius. Put the value of 20 into the formula for volume.

$$20 = \tfrac{4}{3}\pi r^3$$

$$r^3 = \frac{3 \times 20}{4\pi}$$

Hence

$$r = \sqrt[3]{\frac{60}{4\pi}} = 1.684$$

Now use this value in the formula for surface area.

$$A = 4\pi \times 1.684^2 = 35.6$$

The surface area is 35.6 cm^2.

Calculator tip:
Having found the value of r, put it in the memory of your calculator.
Use this saved value for the calculation of the surface area, rather than the rounded value of 1.684.
This is more accurate, and you don't have to key in the value.
Write down 1.684, but use the saved value.

Exercise 3.3

1 Find the volume of a sphere with
  a radius 10 cm   **b** radius 6 cm   **c** diameter 28 cm.
2 Find the surface areas of the spheres of question 1.
3 Assuming that the Earth is a sphere of radius 6400 km, find its volume and surface area.
4 The volume of a sphere is 60 cm^3. Find its radius.
5 The volume of a sphere is 0.7 m^3. Find its radius.
6 The surface area of a sphere is 21 m^2. Find its radius.
7 The surface area of a sphere is 88 mm^2. Find its radius.
8 The volume of a sphere is 34 m^3. Find its surface area.
9 The surface area of a sphere is 70 mm^2. Find its volume.
10 Complete this table for three spheres.

	radius	surface area	volume
sphere A	3 cm		
sphere B		0.8 m^2	
sphere C			97 m^3

11 A sphere of radius 10 cm is to be covered with gold leaf, 0.001 cm thick. Find the volume of gold used, in the following two ways, and check that the answers are approximately equal.
  a Subtract the volume of the sphere before covering, from the volume after covering.
  b Multiply the surface area of the sphere by the thickness of the gold leaf.
12 A sphere of radius 12 cm is cut into two hemispheres. Find the total surface area of each hemisphere.
13 Find the total surface area of a hemisphere of radius r cm.

Dimensions

Look at these incorrect statements.

> He ran a length of 2.3 km^2.
> The area of the floor is 16.3 m^3.
> The pond contains 183 m of water.

All these are wrong because they have the wrong units. A length cannot be measured in km^2, an area cannot be measured in m^3 and a volume cannot be measured in m.

> Length is measured in km (or m, or cm, etc.).
> Area is measured in m^2 (or km^2, or mm^2, etc.).
> Volume is measured in m^3 (or cm^3, or mm^3, etc.).

Length, area and volume have different **dimensions**.

> A length is one-dimensional. It consists of one length.
> An area is two-dimensional. It is found by multiplying two lengths together.
> A volume is three-dimensional. It is found by multiplying three lengths together.

Let us look at some of the formulae we have for lengths, areas and volumes.

Rectangle
The area of a rectangle is found by $l \times b$.
This involves two lengths multiplied together.

Cuboid

The volume of a cuboid is found by *lbh*.

This involves three lengths multiplied together.

Square

The area of a square is found by x^2.

This is $x \times x$, so it involves two lengths multiplied together.

Exercise 3.4

In the following, x, y and z refer to lengths. In each case write down whether the expression could refer to length, area or volume.

1 xy **2** xyz **3** y^2 **4** xz **5** x^3

6 y^2z **7** zx **8** x^2 **9** $\dfrac{xy}{z}$ **10** $\dfrac{x^3}{y}$

Many formulae involve numbers as well as lengths. For example:

Triangle

The area of a triangle is found by $\frac{1}{2}bh$.

Circle

The area of a circle is found by πr^2.

The circumference of a circle is found by $2\pi r$.

Cylinder

The volume of a cylinder is found by $\pi r^2 h$.

The terms $\frac{1}{2}$, π and 2 are just numbers. They do not represent length, area or volume. They have no dimensions, so they do not affect the dimensions of the formula.

> $\frac{1}{2}bh$ and πr^2 each involve two lengths multiplied together, which gives area.
> $2\pi r$ involves a single length.
> $\pi r^2 h$ involves three lengths multiplied together, which gives volume.

Example   In the expression $2\pi lb$, l and b refer to lengths. The numbers 2 and π have no dimensions. Does the expression refer to length, area or volume?

The numbers 2 and π don't affect the dimensions of the formula. Remove them, obtaining *lb*. This involves two lengths multiplied together.

    The expression could refer to an area.

Exercise 3.5

In the following, x, y and z are lengths. The numbers have no dimensions. In each case write down whether the expression could refer to length, area or volume.

1 $5x$ **2** $\frac{1}{4}y^2$ **3** $3xyz$ **4** $\frac{4}{3}\pi x^3$

5 πxy **6** $3\pi z$ **7** $5x^2y$ **8** $0.3xz^2$

9 $\dfrac{\pi x^3}{2y^2}$ **10** $\dfrac{17x^2y}{\pi z}$

Some expressions involve lengths added together. The result is still a length. Think of two sticks laid end to end – the total length is still a length.

Perimeter of a rectangle

The perimeter of a rectangle is found by $2l + 2b$. The 2 is just a number, with no dimensions. So the expression gives us two lengths added together, which is another length.

> **Perimeter:** from the Greek *peri* meaning 'around' and *metron* meaning 'measure'.

Example   In the expression $\pi ab + \pi ac$, a, b and c refer to lengths. π is a number with no dimensions. Which of length, area or volume could the expression refer to?

As π is just a number, remove it, obtaining $ab + ac$. Both ab and ac refer to area. The sum of two areas is another area.

The expression could refer to an area.

Note. A length added to a length is another length. But it does not make sense to add a length and an area. Terms with different dimensions cannot be added or subtracted.

Exercise 3.6

In questions 1–14, x, y and z are lengths. The numbers have no dimensions. Write down whether each expression could refer to length, area or volume.

1 $x + y$

2 $x^2 + xy$

3 $z^3 - xyz$

4 $5x + 3y$

5 $\pi x^2 - 2yz$

6 $2\pi y + \pi z$

7 $\frac{1}{3}\pi x^3 + 4xyz$

8 $\pi(x + y)$

9 $3x(x + y)$

10 $xz + zy + yx$

11 $z(x^2 + y^2)$

12 $(x + y)(x + z)$

13 $2\pi x(x + y)(x + z)$

14 $4\pi x^2 + 2\pi x(x + y)$

15 A cuboid is a by b by c. Which of length, area or volume do the following refer to?

   a abc **b** $4a + 4b + 4c$ **c** $2(ab + bc + ca)$

16 A cylinder has height h and base radius r. Which of length, area or volume do the following refer to?

   a $2\pi r(r + h)$ **b** $4\pi r$ **c** $\pi r^2 h$

17 A cone has base radius r and height h. Which of length, area or volume do the following refer to?

   a $\pi r^2 h$ **b** $2\pi r$ **c** $\sqrt{r^2 + h^2}$ **d** $\pi r\sqrt{r^2 + h^2}$

18 The letters a and b refer to length.

   a The formula $a^n b$ refers to volume. Find n.

   b The formula $a^n(a + b)$ refers to area. Find n.

SUMMARY

■ The volume of a **prism** is the product of the area of cross-section and the height, $V = Ah$. For example, if $A = 8\,cm^2$ and $h = 3\,cm$ then $V = 24\,cm^3$.

■ **Pyramids** and **cones** slope uniformly to a point from a base. Their volume is a third the base area times the height, $V = \frac{1}{3}Ah$. For example, if a cone has a radius $2\,m$ and height $4\,m$ its volume is $\frac{1}{3}\pi 2^2 4\,m^3$, which is $16.8\,m^3$.

■ If a **sphere** has a radius r, its volume is $\frac{4}{3}\pi r^3$ and its surface area is $4\pi r^2$. For example, if a sphere has a radius $5\,m$, its volume is $\frac{4}{3}\pi 5^3\,m^3$, which is $524\,m^3$. Its surface area is $4\pi 5^2\,m^2$, which is $314\,m^2$.

■ Length, area and volume have **dimensions** of length, length squared and length cubed respectively. You can tell which of length, area or volume a formula represents.

Exercise 3A

1 The cross-section of a prism is a triangle with sides $10\,cm$, $8\,cm$ and $6\,cm$. The prism is $12\,cm$ long. Find its volume.

2 A cylinder has height $0.36\,m$ and volume $9.2\,m^3$. Find its base radius.

3 Find the volume of a cone which has base radius $4\,cm$ and height $12\,cm$.

4 A cone has base radius $4\,cm$ and volume $32\,cm^3$. Find its height.

5 A pyramid with a square base has volume $12\,m^3$ and height $4\,m$. Find the side of its base.

6 A sphere has radius $5\,cm$. Find its volume.

7 Find the surface area of the sphere of question 6.

8 The surface area of a sphere is $66\,mm^2$. Find its radius.

9 Find the volume of the sphere of question 8.

10 An **ellipse** is an oval shape. The greatest and least lengths from the centre to the circumference are a and b respectively.

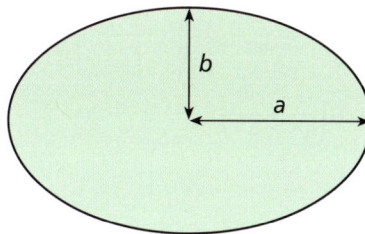

Which of length, area or volume do the following refer to?
   a $\pi(a + b)$ **b** πab **c** $\frac{4}{3}\pi a^2 b$

Exercise 3B

1 Find the volume of a cylinder which has base radius $12\,cm$ and height $40\,cm$.

2 The cross-section of a prism is a right-angled triangle with sides $5\,cm$, $12\,cm$ and $13\,cm$. The length of the prism is $14\,cm$. Find its volume.

3 Find the surface area of the prism of question 2.

4 A pyramid has a rectangular base which is $12\,m$ by $15\,m$. Its height is $20\,m$. Find its volume.

5 Find the volume of a cone which has base radius $1.5\,m$ and height $0.6\,m$.

6 Find the base radius of a cone which has height $12\,cm$ and volume $100\,cm^3$.

7 Find the volume of a sphere with radius $1.6\,mm$.

8 The volume of a sphere is $87\,mm^3$. Find its radius.

9 Find the surface area of the sphere of question 8.

10 The diagram shows a cone with its top cut off (called a **frustum**). It has height h, base radius R and top radius r.

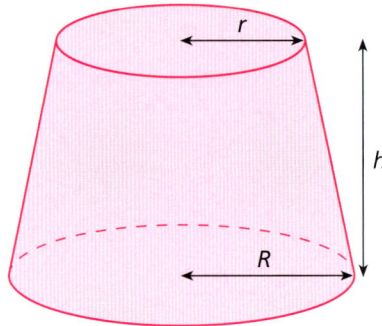

Which of length, area or volume do the following refer to?
a $\frac{1}{3}\pi h(r^2 + Rr + R^2)$ **b** $\sqrt{h^2 + (R-r)^2}$ **c** $\pi(R+r)\sqrt{h^2 + (R-r)^2}$

Exercise 3C

1 The method of dimensions can be used to check whether a formula is correct. What is wrong with the following formulae? Here x, y and z refer to lengths.

$x + y^2$ $(xy + z^3)$ $2x + yz$

2 The method of dimensions can be extended to include mass and time. Which of the following formulae *must* be wrong? Here M refers to a mass, and t and T refer to times.

$M(x+t)$ $\dfrac{(x+y)}{t}$ $\dfrac{Mx}{t^2} - \dfrac{My}{T^2}$ $\dfrac{Mx}{t} + \dfrac{MT}{y}$

$x^2t - My^2T$ $3\pi xt + yT$ $M\left(\dfrac{1}{y^2} + \dfrac{1}{x^2}\right)$ $\frac{1}{4}x^2 + txy$

Exercise 3D

These problems are often best solved in a group.

1 Suppose you have an open cubical box which can hold exactly 6 litres of water. You can use it to take water from a vat once and once only.

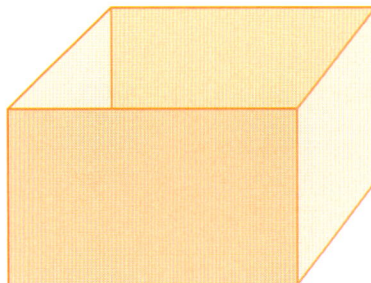

Show how you can use the box to measure out exactly
a 3 litres   **b** 1 litre   **c** 2 litres.

2 This diagram shows an open box with the sides as given.

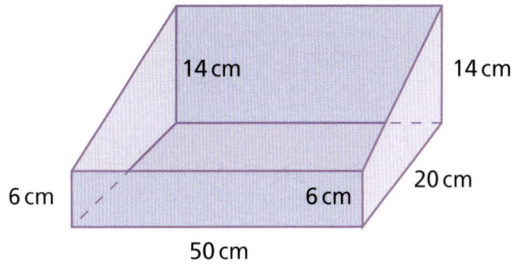

Again, you can use the box to take water from a vat once and once only. Using this box, what volumes can you measure out exactly?

Exercise 3E Ma1

The volume of a sphere is given by the formula $\frac{4}{3}\pi r^3$. Where does this formula come from? Here is an outline of the proof.

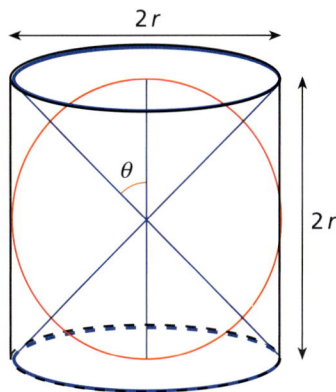

In the diagram a sphere of radius r is enclosed in a cylinder. A 'double cone' is also inside the cylinder. You will show that

volume of sphere = volume of cylinder − volume of cone

1 What are the height and radius of the cylinder?
2 What are the height and radius of each of the cones?
3 What is the angle θ between the slant height of the cone and the height?

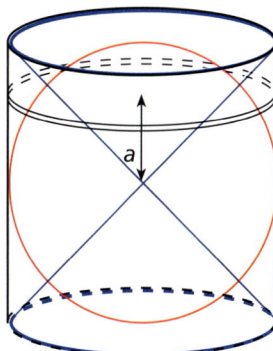

Take a thin horizontal slice through the cylinder, as shown. The slice is a above the centre of the sphere.

The top of the slice will cut the cylinder, cone and sphere in three circles.

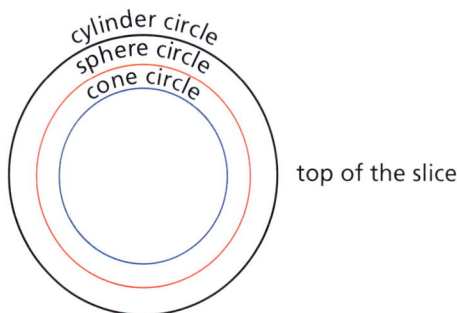

cylinder circle
sphere circle
cone circle

top of the slice

4 What is the radius of the circle where the slice cuts the cylinder?

5 What is the radius of the circle where the slice cuts the cone? Remember your answer to part 3.

6 What is the radius of the circle where the slice cuts the sphere? You will have to use Pythagoras' theorem for this.

7 Find the areas of the three circles. You should find that

> Remember:
>
> *Pythagoras' theorem*
> $c^2 = a^2 + b^2$

    area of sphere circle = area of cylinder circle − area of cone circle

As the slice is thin, the volume of the slice is proportional to the area of its top. Add up all the slices from the top to the bottom of the diagram. These give the volumes of the solids. Hence

    volume of sphere = volume of cylinder − volume of cone

8 Use the known formulae for volumes of a cylinder and a cone to find the volume of the sphere.

*This was just an outline of the proof. In particular, we skated over the step after 7, of adding up volumes of thin slices to find the total volume. If you go on with mathematics after GCSE, you will meet a systematic method of adding up thin slices to make volumes (part of **integral calculus**).*

4 Quadratics

Expansion

When two pairs of brackets are multiplied together, *all* the terms in the first pair must multiply *all* the terms in the second pair. This is called **expansion**.

$$(a + b)(x + y) = ax + ay + bx + by$$

The diagram shows a rectangle with sides $(a + b)$ and $(x + y)$. Notice that it is made up of four smaller rectangles, with areas ax, ay, bx and by. This verifies the expansion formula.

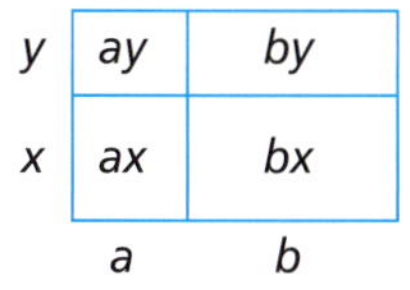

y	ay	by
x	ax	bx
	a	b

After expansion, some expressions can be simplified by collecting like terms.

Example Expand and simplify $(2x + 3)(3x - 5)$.

Both terms in the first pair of brackets are multiplied by both terms in the second pair.

$$(2x + 3)(3x - 5) = 2x \times 3x + 2x \times (-5) + 3 \times 3x + 3 \times (-5)$$
$$= 6x^2 - 10x + 9x - 15$$
$$= 6x^2 - x - 15$$

Exercise 4.1

Expand and simplify the following.

1 $(x + 7)(x + 3)$
2 $(x + 8)(x + 4)$
3 $(x - 4)(x - 2)$
4 $(x - 5)(x - 6)$
5 $(x + 3)(x - 5)$
6 $(x - 2)(x + 7)$
7 $(2x + 3)(x + 4)$
8 $(3x + 1)(x + 5)$
9 $(3x - 7)(x + 1)$
10 $(x + 3)(4x - 5)$
11 $(x - 5)(4x - 3)$
12 $(2x - 3)(x - 5)$
13 $(3x + 1)(2x + 7)$
14 $(5x + 1)(4x + 1)$
15 $(4x - 3)(3x + 9)$
16 $(2x + 5)(3x - 2)$
17 $(2x - 7)(5x - 2)$
18 $(7x - 3)(2x - 3)$
19 $(\frac{1}{2}x + 9)(\frac{1}{3}x + 12)$
20 $(\frac{1}{4}x + 2)(\frac{1}{2}x - 3)$
21 $(\frac{1}{8}x - 3)(\frac{2}{3}x - 4)$

Factorisation

Expansion gets rid of brackets. The opposite procedure is **factorisation**. This puts an expression into brackets.

Consider the expansion of $(x + 3)(x + 2)$.

$$(x + 3)(x + 2) = x^2 + 3x + 2x + 6 = x^2 + 5x + 6$$

Notice that the x term, $5x$, was found by adding $3x$ and $2x$. So the coefficient of x is the *sum* of 3 and 2.

Notice that the constant term, 6, was found by multiplying 3 and 2. So the constant term is the *product* of 3 and 2.

> Notice that the order of terms here is different from above. This doesn't matter and will not affect the solution.

This is true in general. When factorising a **quadratic expression** like $x^2 + bx + c$, look for two numbers whose product is c and whose sum is b. So find the factors of c, and see which pair has a sum of b.

Example   Factorise $x^2 + 7x + 10$.

If this is written $(x + a)(x + b)$, we want a and b such that $ab = 10$ and $a + b = 7$. The factors of 10 are

$$1 \times 10 \qquad 2 \times 5$$

The sum of 2 and 5 is 7. So this is the pair we want.
   Hence $x^2 + 7x + 10 = (x + 2)(x + 5)$.

Exercise 4.2

Factorise these expressions.

1 $x^2 + 4x + 3$	**2** $x^2 + 5x + 6$	**3** $x^2 + 8x + 15$
4 $x^2 + 5x + 4$	**5** $x^2 + 7x + 12$	**6** $x^2 + 9x + 18$
7 $x^2 + 8x + 12$	**8** $x^2 + 10x + 21$	**9** $x^2 + 11x + 30$
10 $x^2 + 11x + 18$	**11** $x^2 + 12x + 20$	**12** $x^2 + 11x + 28$
13 $x^2 + 13x + 30$	**14** $x^2 + 11x + 24$	**15** $x^2 + 13x + 40$
16 $x^2 + 13x + 42$	**17** $x^2 + 17x + 72$	**18** $x^2 + 15x + 54$
19 $x^2 + 24x + 63$	**20** $x^2 + 25x + 150$	

Consider the expansion of $(x - 3)(x - 4)$.

$$-3 \times -4 = +12$$

$$(x - 3)(x - 4) = x^2 - 3x - 4x + 12$$
$$= x^2 - 7x + 12$$

Notice that the sum of -3 and -4 is -7, and that the product of -3 and -4 is $+12$. So we have factors of 12 whose sum is -7. Both these factors are negative.

Example   Factorise $x^2 - 5x + 6$.

If this is written as $(x + a)(x + b)$, we want a and b such that $ab = 6$ and $a + b = -5$. So we want factors of 6 whose sum is -5. This means both these factors must be negative. The (negative) factors of 6 are

$$-1 \times -6 \qquad -2 \times -3$$

Of these, the sum of -2 and -3 is -5.
   So $x^2 - 5x + 6 = (x - 2)(x - 3)$.

Exercise 4.3

Factorise these expressions.

1 $x^2 - 6x + 8$	**2** $x^2 - 3x + 2$	**3** $x^2 - 7x + 6$
4 $x^2 - 9x + 14$	**5** $x^2 - 8x + 12$	**6** $x^2 - 9x + 20$
7 $x^2 - 17x + 60$	**8** $x^2 - 6x + 5$	**9** $x^2 - 12x + 27$
10 $x^2 - 12x + 32$	**11** $x^2 - 14x + 48$	**12** $x^2 - 16x + 63$
13 $x^2 - 18x + 80$	**14** $x^2 - 25x + 156$	**15** $x^2 - 29x + 120$

Consider these expansions of $(x+5)(x-2)$ and $(x-5)(x+2)$.

$$(x+5)(x-2) = x^2 + 5x - 2x - 10 = x^2 + 3x - 10$$
$$(x-5)(x+2) = x^2 - 5x + 2x - 10 = x^2 - 3x - 10$$

Notice that in both cases the product of 5 and 2 is 10. But the coefficient of x, $+3$ or -3, is the *difference* of 5 and 2. Put another way, it is the sum of $+5$ and -2, or of -5 and $+2$. So, if the constant term (the end term) is negative, look for factors whose difference is the coefficient of x (the middle term). If this middle term is positive, make the larger factor positive and the smaller factor negative. If the middle term is negative, make the larger factor negative and the smaller factor positive.

Examples   Factorise $x^2 - 7x - 18$.

The factors of 18 are 1×18, 2×9 and 3×6. Of these, the difference of 2 and 9 is 7.
  Because the coefficient of x is -7, take -9 and $+2$.
  So $x^2 - 7x - 18 = (x-9)(x+2)$.

Factorise $x^2 + x - 20$.

The factors of 20 are 1×20, 2×10 and 4×5. Of these, the difference of 4 and 5 is 1.
  Because the coefficient of x is $+1$, take $+5$ and -4.
  So $x^2 + x - 20 = (x+5)(x-4)$.

Exercise 4.4

Factorise these expressions.

1 $x^2 - 2x - 15$
2 $x^2 - 5x - 14$
3 $x^2 + x - 6$
4 $x^2 + 4x - 21$
5 $x^2 - 2x - 35$
6 $x^2 - 5x - 50$
7 $x^2 - 2x - 48$
8 $x^2 - 4x - 32$
9 $x^2 + 5x - 24$
10 $x^2 + 3x - 70$
11 $x^2 + 6x - 16$
12 $x^2 + 5x - 6$
13 $x^2 - 7x - 30$
14 $x^2 + 17x - 60$
15 $x^2 - 2x - 120$
16 $x^2 + 3x - 40$
17 $x^2 + 3x - 54$
18 $x^2 - 2x - 80$
19 $x^2 - x - 72$
20 $x^2 + 14x - 51$

Here is a general plan for factorising an expression of the form $x^2 + bx + c$. You might find it helpful to follow through this 'questionnaire'.

1 Is c positive? If $c>0$, go to step 2. If $c<0$, go to step 5.
2 Find factors of c whose sum is b.
3 Is $b>0$? If $b>0$, make both factors positive. If $b<0$, make both factors negative.
4 Write out the factorisation. End.
5 Find factors of c whose difference is b.
6 Is $b>0$? If $b>0$, make the larger factor positive and the smaller negative.
   If $b<0$, make the larger factor negative and the smaller positive.
7 Write out the factorisation. End.

In the following exercise the various types of expression are jumbled up.

Exercise 4.5

Factorise the expressions in questions 1–18.

1 $x^2 - 9x + 8$	**2** $x^2 - 7x - 8$	**3** $x^2 + 8x - 9$
4 $x^2 + 14x + 45$	**5** $x^2 - 5x - 36$	**6** $x^2 + 3x - 108$
7 $x^2 + 14x + 40$	**8** $x^2 + 2x - 63$	**9** $x^2 - 15x + 44$
10 $x^2 - 4x - 45$	**11** $x^2 + 9x + 18$	**12** $x^2 + 3x - 180$
13 $x^2 - 120x + 999$	**14** $x^2 - 6x - 40$	**15** $x^2 - 23x + 132$
16 $x^2 + 22x + 120$	**17** $x^2 - 90x - 1000$	**18** $x^2 - 25x + 100$

19 Suppose p is a prime number, and $x^2 + kx + p$ factorises. What is k?

20 Suppose p is a prime number, and $x^2 + kx - p$ factorises. What is k?

Three special factorisations

Consider these three expansions.

$$(x + a)^2 = (x + a)(x + a) = x^2 + 2ax + a^2$$
$$(x - a)^2 = (x - a)(x - a) = x^2 - 2ax + a^2$$
$$(x + a)(x - a) = x^2 - a^2$$

> This is known as the **difference of two squares**.

Clearly each of these expansions gives a corresponding factorisation. In particular, any expression which consists of one square subtracted from another is a 'difference of two squares'. Examples are

$$a^2 - b^2 \qquad x^2 - 100 \qquad 4x^2 - 1$$

Examples   Factorise $x^2 - 6x + 9$.

This is of the form $(x - a)^2 = x^2 - 2ax + a^2$, with $a = 3$.
So $x^2 - 6x + 9 = (x - 3)^2$.

Factorise $x^2 - 49$.

This is of the form $(x + a)(x - a) = x^2 - a^2$, with $a = 7$.
So $x^2 - 49 = (x + 7)(x - 7)$.

Exercise 4.6

Expand the expressions in questions 1–6.

1 $(x + 3)^2$	**2** $(x + 7)^2$	**3** $(x - 4)^2$
4 $(x - 5)^2$	**5** $(x + 1.5)^2$	**6** $(x - \frac{1}{4})^2$

Factorise the expressions in questions 7–21.

7 $x^2 + 4x + 4$	**8** $x^2 - 2x + 1$	**9** $x^2 + 8x + 16$
10 $x^2 + 12x + 36$	**11** $x^2 - 9$	**12** $x^2 + 10x + 25$
13 $x^2 - 16x + 64$	**14** $x^2 - 100$	**15** $x^2 - 18x + 81$
16 $x^2 - 64$	**17** $x^2 - \frac{1}{4}$	**18** $4 - x^2$
19 $x^2 - 20x + 100$	**20** $x^2 - x + \frac{1}{4}$	**21** $x^2 - 2.25$

22 $x^2 + kx + m$ factorises to $(x + a)^2$, for some a. Write m in terms of k.

Quadratic equations

Suppose a quadratic expression is made equal to 0. We get an equation of the form $x^2 + bx + c = 0$. This is a **quadratic equation**. This sort of equation is more complicated than a linear equation.

The graph of $y = x^2 + bx + c$ looks like one of the three pictures shown here. The solutions to the corresponding quadratic equation are found where the graph meets the x-axis. As you can see, the graph can meet the axis at two points, one point or no point at all.

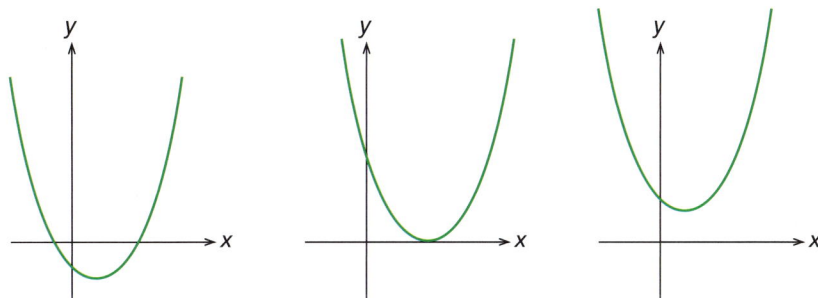

Hence there may be two solutions to the quadratic equation, there may be one solution, or there may be no solutions at all. One way to solve a quadratic equation is by factorising the left-hand side.

Suppose the left-hand side is factorised to $(x + m)(x + n)$. The equation becomes

$$(x + m)(x + n) = 0$$

Now, if the product of two terms is 0, then one of the terms must be 0. (If both terms were non-zero, then their product would also be non-zero.) So

either $x + m = 0$ or $x + n = 0$
either $x = -m$ or $x = -n$

Example   Solve the equation $x^2 + 5x - 24 = 0$.

Factorise the left-hand side, to $(x + 8)(x - 3)$. The equation is

$$(x + 8)(x - 3) = 0$$

Either $x + 8 = 0$ or $x - 3 = 0$. The first gives $x = -8$, the second gives $x = 3$.
So $x = -8$ or $x = 3$.

Exercise 4.7

Solve the equations in questions 1–17.

1 $x^2 - 5x + 6 = 0$
2 $x^2 - 8x + 15 = 0$
3 $x^2 + 6x + 8 = 0$
4 $x^2 + 11x + 30 = 0$
5 $x^2 - 2x - 35 = 0$
6 $x^2 + 4x - 12 = 0$
7 $x^2 + 7x - 44 = 0$
8 $x^2 - x - 30 = 0$
9 $x^2 - 12x + 36 = 0$
10 $x^2 - 6x + 9 = 0$
11 $x^2 + 18x + 81 = 0$
12 $x^2 + \frac{2}{3}x + \frac{1}{9} = 0$
13 $x^2 + x + \frac{1}{4} = 0$
14 $x^2 - 64 = 0$
15 $x^2 - 10000 = 0$
16 $x^2 - \frac{1}{4} = 0$
17 $0.01 - x^2 = 0$

18 p is a prime, and the expression $x^2 + kx - p$ factorises, where k is a positive constant. Solve the equation $x^2 + kx - p = 0$.

19 p is a prime, and the expression $x^2 - kx + p$ factorises, where k is a positive constant. Solve the equation $x^2 - kx + p = 0$.

The expression $ax^2 + bx + c$

A general quadratic expression is of the form $ax^2 + bx + c$. The quadratics on page 45 all began with x^2, so they all had $a = 1$.

Consider the expansion of $(6x + 5)(3x + 4)$.

$$(6x + 5)(3x + 4) = 18x^2 + 24x + 15x + 20$$
$$= 18x^2 + 39x + 20$$

Note that 18 is the product of 6 and 3, and that 20 is the product of 5 and 4. The coefficient of x, 39, is $6 \times 4 + 5 \times 3$. So if we were asked to factorise $18x^2 + 39x + 20$, there would be many possible factors of 18 and 20 that we would have to try. The following is a systematic approach to factorising $ax^2 + bx + c$.

● *To factorise $ax^2 + bx + c$, first form the product ac. Find factors of ac whose sum is b. Rewrite the quadratic with the bx term expressed as a sum of these factors.*

   You now have four terms. Group these terms into two pairs and factorise each pair. You will then find that there is another common factor that can be taken out.

Example   Factorise $6x^2 + 19x + 10$.

The product ac is 60. The factors of 60 whose sum is 19 are 4 and 15. Write $19x$ in terms of these numbers, and group into two pairs.

$$6x^2 + 19x + 10 = 6x^2 + 4x + 15x + 10$$
$$= (6x^2 + 4x) + (15x + 10)$$

Take out $2x$ from the first pair, and 5 from the second pair.

$$(6x^2 + 4x) + (15x + 10) = 2x(3x + 2) + 5(3x + 2)$$

Now there is a common factor of $(3x + 2)$. It is multiplied by $2x$ and by 5. Factorise by $(3x + 2)$.

$$6x^2 + 19x + 10 = (3x + 2)(2x + 5)$$

Exercise 4.8

Factorise these expressions.

1 $2x^2 + 5x + 2$ **2** $3x^2 + 8x + 5$ **3** $2x^2 + 15x + 18$
4 $2x^2 + 9x + 7$ **5** $3x^2 + 26x + 35$ **6** $5x^2 + 16x + 3$
7 $5x^2 + 12x + 4$ **8** $4x^2 + 7x + 3$ **9** $4x^2 + 8x + 3$
10 $6x^2 + 17x + 5$ **11** $4x^2 + 11x + 6$ **12** $8x^2 + 10x + 3$
13 $6x^2 + 19x + 8$ **14** $10x^2 + 19x + 6$ **15** $14x^2 + 31x + 15$

Examples Factorise $5x^2 - 11x + 6$.

$5 \times 6 = 30$. We want factors of 30 whose sum is 11. Because the x term is -11, both the factors are negative. They are -5 and -6. When the terms go into brackets, be careful with the negative values. If there is a minus sign outside the brackets, change the sign of the terms inside the brackets.

$$5x^2 - 11x + 6 = 5x^2 - 5x - 6x + 6$$
$$= (5x^2 - 5x) - (6x - 6)$$
$$= 5x(x - 1) - 6(x - 1)$$
$$= (x - 1)(5x - 6)$$

$-6 \times -1 = +6$

$5x^2 - 11x + 6 = (x - 1)(5x - 6)$.

Factorise $12x^2 - x - 6$.

The product of 12 and 6 is 72. Because the constant term is negative, we want factors of 72 whose *difference* is 1. These are 9 and 8. Because the x term is negative, we want -9 and $+8$.

$$12x^2 - x - 6 = 12x^2 - 9x + 8x - 6$$
$$= (12x^2 - 9x) + (8x - 6)$$
$$= 3x(4x - 3) + 2(4x - 3)$$
$$= (4x - 3)(3x + 2)$$

$12x^2 - x - 6 = (4x - 3)(3x + 2)$.

Exercise 4.9

Factorise the expressions in questions 1–14.

1 $2x^2 + 7x - 15$　　**2** $2x^2 - 17x + 30$　　**3** $3x^2 - 22x + 7$
4 $6x^2 + 13x - 5$　　**5** $4x^2 - 16x + 15$　　**6** $4x^2 - 4x - 3$
7 $5x^2 - 12x - 9$　　**8** $7x^2 - 16x + 4$　　**9** $3x^2 - 4x - 15$
10 $6x^2 + 13x - 8$　　**11** $8x^2 - 18x + 9$　　**12** $10x^2 - 43x - 9$
13 $14x^2 + 17x - 6$　　**14** $15x^2 + x - 6$
15 p and q are different prime numbers, and the expression $px^2 + kx + q$ factorises. Find the possible values of k.

If the expression $ax^2 + bx + c$ factorises, then you can solve the equation $ax^2 + bx + c = 0$.

Example Solve the equation $12x^2 - x - 6 = 0$.

The left-hand side has already been factorised in the previous worked example, as $(4x - 3)(3x + 2)$. Hence

$$(4x - 3)(3x + 2) = 0$$

So either $4x - 3 = 0$ or $3x + 2 = 0$.

$$x = \tfrac{3}{4} \text{ or } x = -\tfrac{2}{3}$$

Exercise 4.10

Solve the equations in questions 1–15.

1 $3x^2 + x - 4 = 0$
2 $2x^2 + 11x + 15 = 0$
3 $2x^2 - 7x - 4 = 0$
4 $6x^2 - 19x + 10 = 0$
5 $5x^2 - 12x + 4 = 0$
6 $6x^2 - 11x + 4 = 0$
7 $9x^2 - 6x - 8 = 0$
8 $10x^2 - 31x + 15 = 0$
9 $15x^2 + 37x - 8 = 0$
10 $4x^2 - 4x + 1 = 0$
11 $9x^2 + 6x + 1 = 0$
12 $4x^2 - 12x + 9 = 0$
13 $25x^2 - 20x + 4 = 0$
14 $9x^2 - 4 = 0$
15 $4x^2 - 81 = 0$

16 p and q are different prime numbers, and the expression $px^2 + kx - q$ factorises. Find the possible pairs of solutions of $px^2 + kx - q = 0$.

Sometimes we need to adjust the equation to get it in the form to be factorised. We must collect all the terms on the left-hand side, so that there is only 0 on the right-hand side.

Example Solve these equations.

a $x^2 - 4x = 12$ **b** $x + \dfrac{6}{x} = 5$

a Subtract 12 from each side.

$$x^2 - 4x - 12 = 0$$
$$(x - 6)(x + 2) = 0$$
$$x = 6 \text{ or } x = -2$$

Now there is 0 on the right-hand side.

b Multiply both sides by x.

$$\dfrac{6}{x} \times x = 6$$

$$x^2 + 6 = 5x$$
$$x^2 - 5x + 6 = 0 \quad \text{(subtracting } 5x \text{ from each side)}$$
$$(x - 2)(x - 3) = 0$$
$$x = 2 \text{ or } x = 3$$

Exercise 4.11

Solve these equations.

1 $x^2 + 12 = 7x$
2 $x^2 = 5x - 6$
3 $8 = 2x + x^2$
4 $10 = 7x - x^2$
5 $3x^2 - 2x = 8$
6 $17x = 6 + 5x^2$
7 $(x + 4)(x - 3) = 18$
8 $(x - 3)(x - 5) = 35$
9 $(x + 4)(x - 2) = 7x - 2$
10 $(2x + 1)(x - 5) = 4x + 2$
11 $(3x - 2)(2x - 5) = 2x + 1$
12 $\dfrac{9}{x} - 4x = 9$
13 $x + 3 = \dfrac{40}{x}$
14 $\dfrac{6}{x} + \dfrac{12}{x + 1} = 5$
15 $\dfrac{15}{x + 3} + \dfrac{21}{x + 5} = 6$
16 $\dfrac{2x + 3}{x} = \dfrac{x - 1}{x - 4}$
17 $\dfrac{5x + 2}{3x - 1} = \dfrac{2x + 11}{x + 5}$

Equation problems

Practical problems can sometimes be solved by quadratic equations.

Examples   A field is 10 m longer than it is wide. The area is 7200 m². What is the width?

Let the width be x m. Then the length is $(x + 10)$ m. The product of these gives the area.

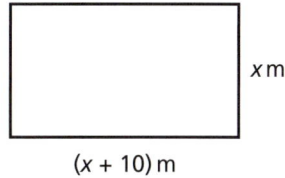

$(x + 10)$ m

$$x(x + 10) = 7200$$
$$x^2 + 10x = 7200$$
$$x^2 + 10x - 7200 = 0$$
$$(x + 90)(x - 80) = 0$$

$x = 80$ or $x = -90$. You cannot have a negative length, so the second answer is meaningless.

The width of the field is 80 m.

Mrs Sinclair drove a distance of 180 miles. Because of traffic, her average speed was 15 m.p.h. less than expected, and the journey took 1 hour longer than expected. Find her average speed.

Suppose her average speed was x m.p.h. Then her expected speed was $(x + 15)$ m.p.h. To find the time taken, divide the distance by the speed.

Remember:

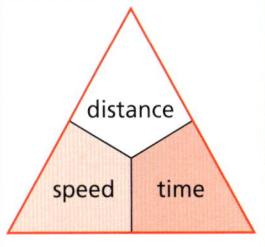

$$\text{time taken} = \frac{180}{x} \text{ hours}$$

$$\text{expected time taken} = \frac{180}{x + 15} \text{ hours}$$

The difference between these times is 1 hour.

$$\frac{180}{x} - \frac{180}{x + 15} = 1$$

Multiply across by x and by $(x + 15)$, then rearrange and solve.

$$180(x + 15) - 180x = x(x + 15)$$
$$180x + 2700 - 180x = x^2 + 15x$$
$$x^2 + 15x - 2700 = 0$$
$$(x - 45)(x + 60) = 0$$

$x = 45$ or $x = -60$. Ignore the negative answer.

Her speed was 45 m.p.h.

Exercise 4.12

1 The side of a square is x cm. The numerical value of the area (in cm^2) added to the numerical value of the perimeter (in cm) is 45. Find x.

2 The sides of a right-angled triangle are x cm, $(2x + 2)$ cm and $(3x - 2)$ cm, where the last of these is the hypotenuse. Find x.

3 A positive number is 42 less than its square. Find the number.

4 The area of a circle (in cm^2) is 35π greater than its circumference (in cm). Find its radius.

5 The two sides containing the right angle of a triangle have sum 23 cm. The area of the triangle is 60 cm^2. Find the sides of the triangle.

6 The height of a cylinder is 5 cm, and its surface area is 132π cm^2. Find the base radius of the cylinder.

> The surface area of a cylinder is given by the formula $2\pi r(r + h)$.

7 A ball is thrown upwards. After t seconds, its height h metres is given by $h = 40t - 5t^2$. Find t when the ball is 75 m high.

8 Mr Jenner walks on a journey of 20 miles and back. His average speed for the return journey was 1 m.p.h. greater than for the outward journey. The total time was 9 hours. Find his speed on the outward journey.

9 The current in a river is 2 km per hour. A man can row at x km per hour. He rows 15 km with the current, then 15 km against the current. The total time taken is 18 hours. Find x.

10 The sum of the ages in a group is 396 years. A person aged 48 joins, and the average increases by one year. How many people were in the group originally?

11 A Japanese tourist changed 240 000¥ to pounds at a rate of x¥ per pound. Next day the rate per pound fell by 10¥, and he would have got £40 more if he had waited until then. Find x.

12 An amount of gas has mass 1.2 kg. If its volume increases by 10 m^3, then its density decreases by 0.01 kg/m^3. Find its original volume.

13 The nineteenth-century mathematician Augustus de Morgan was fond of saying that he was x years old in the year x^2. When was he born?

SUMMARY

■ When two pairs of brackets are multiplied together, all terms in the first pair must multiply all terms in the second pair (**expansion**).

■ **Factorisation** is the opposite procedure to expansion.

■ An expression like $x^2 + bx + c$ is a **quadratic expression**. To factorise it, find numbers whose product is c and whose sum is b. If c is positive and b is negative, then both numbers are negative. If c is negative, then one number is positive and the other negative.

■ When a quadratic expression is put equal to 0, the result is a **quadratic equation**. If the expression factorises, then the equation can be solved.

■ To factorise an expression of the form $ax^2 + bx + c$, find numbers whose product is ac and whose sum is b. Write bx in terms of these two numbers, factorise two pairs of terms and then factorise by a linear expression.

■ Some practical problems can be solved with quadratic equations.

Exercise 4A

1 Expand the expression $(x - 8)(x + 3)$.
2 Factorise the expression $x^2 + 11x + 10$.
3 Factorise the expression $x^2 - 14x + 49$.
4 Solve the equation $x^2 - 11x + 24 = 0$.
5 Factorise the expression $5x^2 - 13x + 6$.
6 Factorise the expression $7x^2 + 19x - 6$.
7 Solve the equation $2x^2 + 9x + 9 = 0$.
8 Solve the equation $2x^2 = 7 - 13x$.
9 The length of a rectangle is 18 m greater than the width, and the area is 88 m². Find the width.
10 The current in a river flows at 1 m/s. A woman can swim at x m/s. She swims 80 m with the current, and then 80 m against the current. The difference between her times is 20 seconds. Form an equation in x and solve it.

Exercise 4B

1 Expand the expression $(3x + 4)(2x - 5)$.
2 Factorise the expression $x^2 - 7x - 60$.
3 Factorise the expression $x^2 - 0.01$.
4 Solve the equation $x^2 + 5x - 14 = 0$.
5 Factorise the expression $8x^2 - 21x + 10$.
6 Factorise the expression $9x^2 - 9x - 4$.
7 Solve the equation $9x^2 + 12x + 4 = 0$.
8 Solve the equation $(2x + 1)(3x - 4) = 7$.

9 Solve the equation $\dfrac{12}{x - 1} - \dfrac{12}{x + 2} = 2$.

10 I bought a number of cakes for £9. If each cake had cost 5p less I would have been able to buy 2 more. How much does a cake cost?

Exercise 4C   Ma1

In this chapter you factorised quadratics by finding the factors of the numbers involved. You probably noticed that some numbers have many more factors than others. 13, for example, has only two factors (1 and 13). 12, however, has six factors (1, 2, 3, 4, 6 and 12).

The number 12 has more factors than any previous number, in other words it sets a new record for the number of factors. The next number with more factors is 24, which has eight factors. Find all the numbers up to 100 which have more factors than any previous number.

Exercise 4D   Ma1

Some quadratics factorise, some do not. There is a condition on a, b and c for a quadratic $ax^2 + bx + c$ to factorise.

1 For some of the quadratics of this chapter, for example from exercise 4.9 questions 1–8, find the value of $b^2 - 4ac$. What do they have in common?
2 If you have found a connection, use it to test whether the following quadratics factorise.
   a $x^2 - 43x + 456$ b $x^2 + 87x + 1851$ c $x^2 - 31x - 462$
   d $x^2 + 37x - 342$ e $3x^2 + 43x + 145$ f $5x^2 + 16x - 144$
3 What happens if $b^2 - 4ac = 0$, and what if $b^2 - 4ac < 0$? Investigate. It will help if you sketch the graphs of the quadratics.

Exercise 4E Ma1

To factorise the quadratic $x^2 + bx + c$, it is necessary to factorise the number c. This is easy if c is small. Just take prime numbers, and see whether they divide into c exactly or not. But if c is large, this may take a long time. Just try factorising 192 079 by this method!

Suppose we want to factorise a large number n. We try to find numbers a and b such that

$$n = (a + b)(a - b) = a^2 - b^2$$

So $a^2 - n = b^2$. Find the smallest number k for which $k^2 \geqslant n$. You can do this by finding $\sqrt{n}$, and taking the integer above it. If $k^2 - n$ is a perfect square, then let $a = k$ and $b = \sqrt{k^2 - n}$. If not, try $(k + 1)^2 - n$. If this is a perfect square let $a = k + 1$; if not, continue with $(k + 2)^2 - n$ and so on.

> **Remember:**
> *The difference of two squares*
> $(x + a)(x - a) = x^2 - a^2$

1   Consider 192 079.
   a   Find $\sqrt{192\,079}$. Let k be the whole number just above it.
   b   Find $k^2 - 192\,079$. Show that this is *not* a perfect square.
   c   Find $(k + 1)^2 - 192\,079$. Show that this *is* a perfect square.
   d   Let $a = k + 1$ and $b = \sqrt{(k + 1)^2 - 192\,079}$. Show that 192 079 can be factorised as

$$192\,079 = (a + b) \times (a - b)$$

2   Use this method to factorise the following.
   a   202 211   **b**   26 069

3   We have assumed that n can be factorised as $n = (a + b)(a - b)$. Show that, if n is odd, and $n = xy$, then we can always find a and b so that $x = a + b$ and $y = a - b$.

This method of factorising was found by Pierre Fermat, who lived in the South of France from 1601 to 1665. Fermat is most famous for *Fermat's Last Theorem*, a conjecture which defied proof until 1994, when it was finally solved by the English mathematician Andrew Wiles. There is an interesting book on the whole story written by Simon Singh (published by Fourth Estate, London), and there is a great deal of information available on the internet.

5 Circles

Tangents

Draw a circle and a straight line. There are three possible cases, as shown in the diagrams.

- The straight line may cross the circle at two separate points.

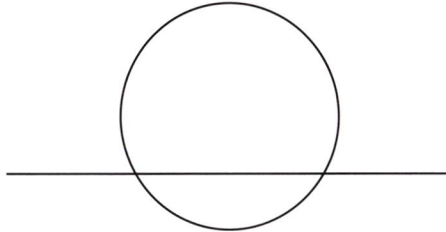

- The line may miss the circle altogether.

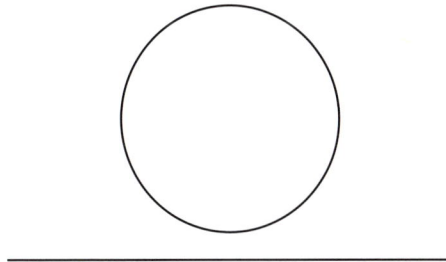

- The line *just* reaches the circle, touching it at only one point.

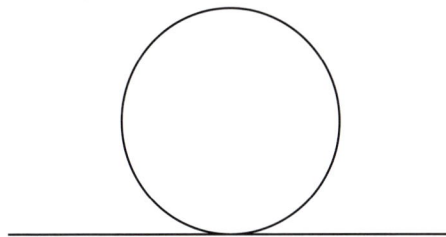

In the third case, the line is called a **tangent** to the circle.

Exercise 5.1

1 Copy this circle. Draw a tangent to the circle.

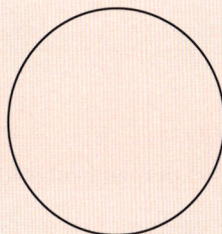

2 This diagram shows two circles. How many lines are there which are tangents to both circles?

These are called **common tangents**.

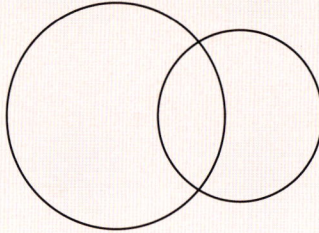

3 Copy the pairs of circles below. For each pair, draw the common tangents.

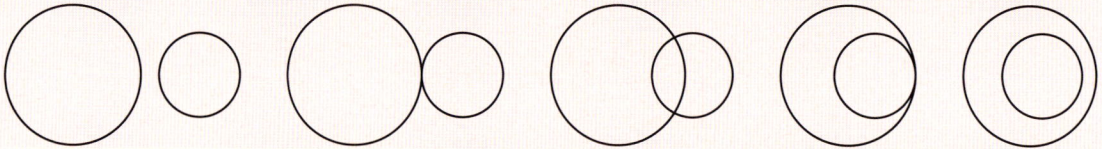

4 Two circles have radii 5 cm and 6 cm. The centres of the circles are d cm apart. How many common tangents are there in each of the following cases?

a $d = 9$ **b** $d = 12$ **c** $d = 11$ **d** $d = 0.5$ **e** $d = 1$

5 Two circles have radii a cm and b cm. The centres of the circles are d cm apart. How many common tangents to the circles are there in each of the following cases?

a $a = 8, b = 6, d = 14$ **b** $a = 9, b = 10, d = 20$ **c** $a = 4, b = 8, d = 3$
d $a = 5, b = 7, d = 2$ **e** $a = 9, b = 8, d = 10$

Two tangent theorems

In the diagram below, the circle has centre O. The tangent meets it at A. Notice that OA is a line of symmetry of the diagram. Hence

$$\angle OAT = \angle OAS = 90°$$

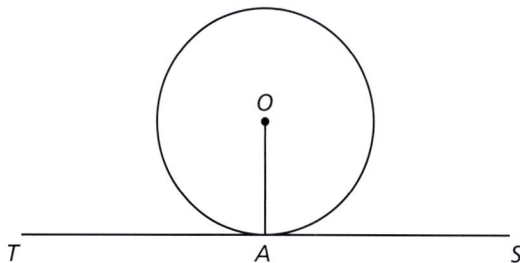

So the radius OA is **perpendicular** to the tangent. This is always true.

● **Theorem 1** *The tangent to a circle is perpendicular to the radius at the point of contact.*

In the diagram below, two tangents are drawn from T, meeting the circle at A and B. Notice that OT is a line of symmetry of the diagram. Hence

$$TA = TB$$

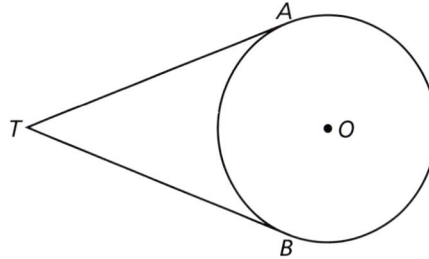

● **Theorem 2** *The tangents to a circle from a point are equal in length.*

Examples   In the diagram below, O is the centre of the circle and TA is a tangent. If $\angle OTA = 35°$ find $\angle TOA$.

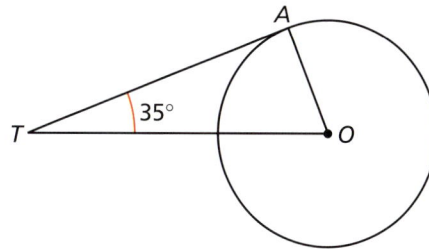

Remember:

If you know two angles of a triangle, you can find the third by subtracting from 180°.

From the result above we know that $\angle OAT = 90°$. Hence

$$\angle TOA = 180° - 90° - 35°$$

$\angle TOA = 55°$.

In the diagram below, TA and TB are the tangents to the circle. What sort of quadrilateral is $TAOB$?

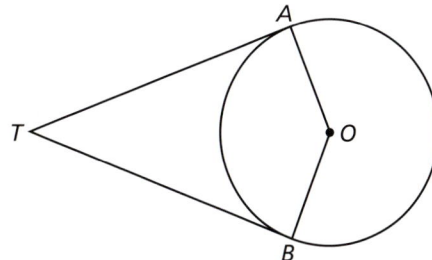

OA and OB are both radii of the circle, hence they are equal. By the result above, $TA = TB$. The quadrilateral has two pairs of adjacent sides equal.
   Hence $TAOB$ is a kite.

Exercise 5.2

1 A circle has centre O. The tangents from T to the circle touch it at A and B. If $\angle AOB = 100°$ find $\angle ATB$.

2 A circle has centre O. A tangent from T meets the circle at A. If $\angle TOA = 66°$ find $\angle OTA$.

3 In this diagram, O is the centre of the circle and TA and TB are tangents. Copy the diagram and mark pairs of equal angles.

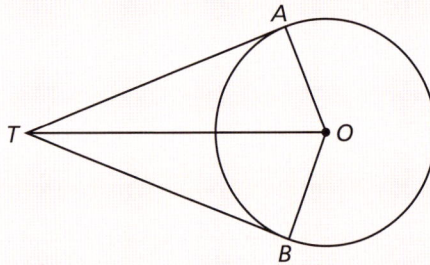

4 The tangents from T to a circle meet it at A and B. If $\angle ATB = 58°$ find $\angle TAB$.

5 The tangents from T to a circle meet it at A and B. If $\angle BAT = 34°$ find $\angle BTA$.

6 In this diagram the circles have centres C_1 and C_2. The line is a tangent to both circles at the same point X. Show that C_1XC_2 is a straight line.

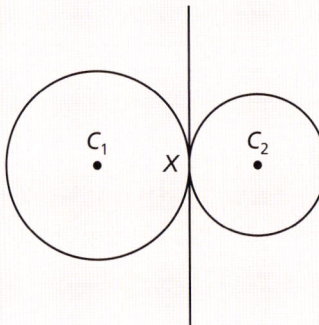

7 O is the centre of a circle, and the tangents from T touch the circle at A and B. If $TAOB$ is a rhombus, what else can you say about $TAOB$?

8 The diagram below shows two circles, with common tangents ATC and BTD. What can you say about AB and CD?

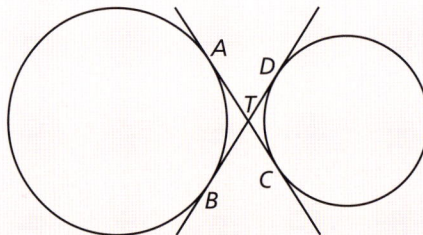

9 Two circles have centres C_1 and C_2 and radii $3\,\text{cm}$ and $5\,\text{cm}$ respectively. Their centres are $2\,\text{cm}$ apart.
  a Show that the circles have one common tangent.
  b Show that C_1, C_2 and the point of contact of the tangent are in a straight line.

Exercise 5.3

It is very difficult to draw an accurate tangent by hand. To construct a tangent at a point on a circle use theorem 1 on page 54.

Method
Suppose we want the tangent at a point A on the circle. Join A to the centre O of the circle. Construct a line making $90°$ with OA. This line is a tangent.

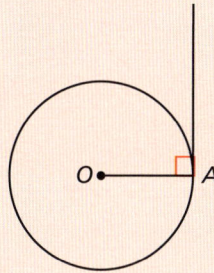

1 Draw a circle with compasses, and construct a tangent by this method. Does it meet the circle only once?
2 Suppose we have two *equal* circles which overlap.

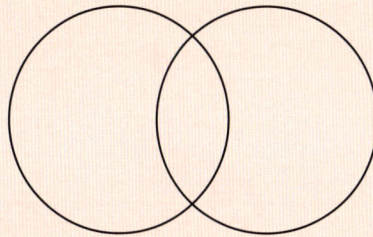

Show how to construct their common tangents.

Remember:
A chord is a line joining two points on a curve.

Perpendicular bisector of a chord

In this diagram, CD is the **perpendicular bisector** of the chord AB. CD goes through the centre of the circle. This is always true.

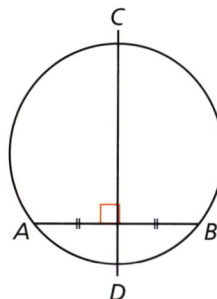

● **Theorem 3** *The perpendicular bisector of a chord of a circle goes through the centre of the circle.*

Proof

Let the midpoint of the chord AB be M, and the centre of the circle be X. in triangles XAM and XBM

$$XA = XB \quad \text{(both radii of circle)}$$
$$AM = BM \quad (M \text{ is the midpoint})$$
$$XM \text{ is common}$$

Hence $\triangle XAM$ and $\triangle XBM$ are congruent (SSS)
Hence $\angle XMA = \angle XMB$, and so $\angle XMA = 90°$.
So XM is the perpendicular bisector of AB.

Example AB is a chord of a circle with centre O. The midpoint of AB is X. If $\angle OAB = 48°$ find $\angle XOA$.

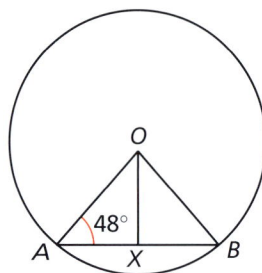

The line XO is perpendicular to AB. Hence $\angle OXA = 90°$. Hence

$$\angle XOA = 180° - 90° - 48°$$

$$\angle XOA = 42°.$$

Exercise 5.4

1 X is the centre of a chord AB of a circle with centre O. If $\angle BOX = 37°$ find $\angle OBX$.
2 M is the centre of a chord PQ of a circle with centre C. If $\angle PQC = 59°$ find $\angle MCQ$.
3 Let the centre of a chord AB be X, and the centre of the circle be O. If $\angle AOX = 62°$ find $\angle OAB$.
4 AB and CD are parallel chords of a circle, with centres X and Y respectively. Show that XY goes through the centre of the circle.
5 A circle has centre O. AB and BC are chords at right angles to each other. If X and Y are the midpoints of AB and BC respectively, show that $XAYO$ is a rectangle.
6 A circle has centre O. AB and BC are equal chords. If X and Y are the midpoints of AB and BC respectively, show that $XBYO$ is a kite.

Exercise 5.5

Suppose you want to find the centre of a circle (perhaps you have drawn the circle by tracing round a circular object, instead of using compasses). Use theorem 3.

Method
Draw two chords *AB* and *CD*. Construct their perpendicular bisectors. These bisectors meet at the centre of the circle.

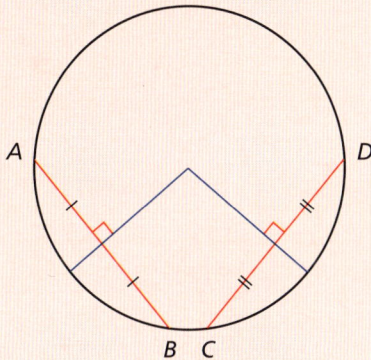

Remember:
To construct the perpendicular bisector of a line segment XY, use compasses to draw an arc from X. With the same separation of the compasses, draw an arc from Y. If the two arcs meet at P and Q, then PQ is the perpendicular bisector of XY.

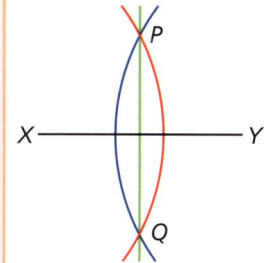

1 Draw a circle by tracing round the base of a mug. Find the centre of the circle by the method above.
2 Is this the true centre of the circle? Test using compasses.

Angle in a semicircle

Suppose *AB* is a chord or an arc of a circle, and *C* is any other point on the circumference. Then we say that $\angle ACB$ is the angle **subtended** at *C* by *AB*.

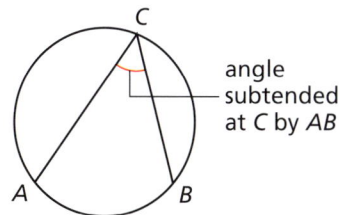

angle subtended at C by AB

If *AB* is the diameter of the circle, and *C* is any other point on the circumference, then the angle subtended by *AB* at *C* is a right angle. This is always true.

● ***Theorem 4*** *Suppose AB is a diameter of a circle, and C is a point on the circumference. Then AB **subtends** a right angle at C, that is, $\angle ACB = 90°$.*

Because AB is a diameter, cutting the circle in half, the angle subtended at C is often called an **angle in a semicircle**.

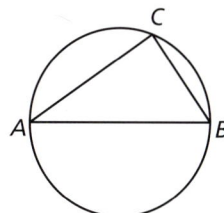

Proof

Let O be the centre of the circle. Then $OA = OB = OC$.
It follows that $\angle OCA = \angle OAC$ (isosceles triangle) and that $\angle OCB = \angle OBC$.

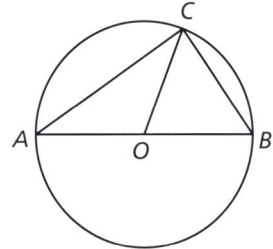

> $\angle COB$ is an exterior angle of $\triangle OCA$.
> $\angle COA$ is an exterior angle of $\triangle OCB$.

$$\angle COB = 2 \times \angle OCA$$
$$\angle COA = 2 \times \angle OCB$$

> **Remember:**
> *Exterior angle =*
> *$180°$ − interior angle*

But $\angle COB + \angle COA = 180°$. It follows that

$$\angle ACB = \angle OCA + \angle OCB = \tfrac{1}{2} \times 180° = 90°$$

Example AB is a diameter of a circle, and C is a point on the circumference. If $\angle BAC = 38°$ find $\angle CBA$.

From the result above, $\angle ACB = 90°$. Hence

$$\angle CBA = 180° - 90° - 38°$$

$$\angle CBA = 52°.$$

Exercise 5.6

1 AB is a diameter of a circle, and C is a point on the circumference. If $\angle ABC = 48°$ find $\angle BAC$.
2 XY is a diameter of a circle, and Z is a point on the circumference. If $\angle ZXY = 57°$ find $\angle ZYX$.
3 PQ is a diameter of a circle with centre C. R is a point on the circumference. If $\angle CRQ = 32°$ find $\angle CPR$.
4 AB is a diameter of a circle with centre X, and C is a point on the circumference. If $\angle AXC = 41°$ find $\angle ABC$.
5 XY is a diameter of a circle with centre O, and Z is a point on the circumference. If $\angle ZOY = 114°$ find $\angle ZXY$.
6 AB is a diameter of a circle and C is a point on the circumference so that $CA = CB$. Find $\angle CAB$.
7 AB is a diameter of a circle, and C is a point on the circumference such that CA is equal to the radius of the circle. Find $\angle CBA$.
8 AB and CD are diameters of a circle. Show that $ACBD$ is a rectangle.
9 You are half-way up a ladder leaning against a wall. The ladder starts to slip down. What path do your feet follow?

Exercise 5.7

In exercise 5.3 you constructed the tangent *at* a point on a circle. How do you construct a tangent *from* a point to a circle?

Method
Join the point T to the centre O of the circle. Find the midpoint C of OT. Draw the circle centre C which goes through T and O. If this circle cuts the original circle at A and B, then TA and TB are the tangents to the circle.

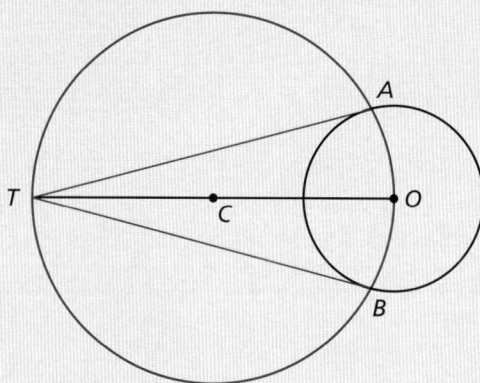

1 Draw a circle, and pick a point outside it. By the method above, construct a tangent from the point to the circle. Does the tangent meet the circle once only?
2 Explain why this method works. (Hint: use theorem 4 on page 59.)

Suppose you have two circles. Here is how to construct the common tangents (if they exist).

Method
Let the circles have centres A and B and radii a and b respectively. Suppose $a > b$. Find $a - b$, and draw a circle centre A with radius $a - b$. Construct a tangent from B to the new circle, meeting it at C. Extend AC to the first circle, cutting it at D. Construct the tangent at D. This will be a common tangent.

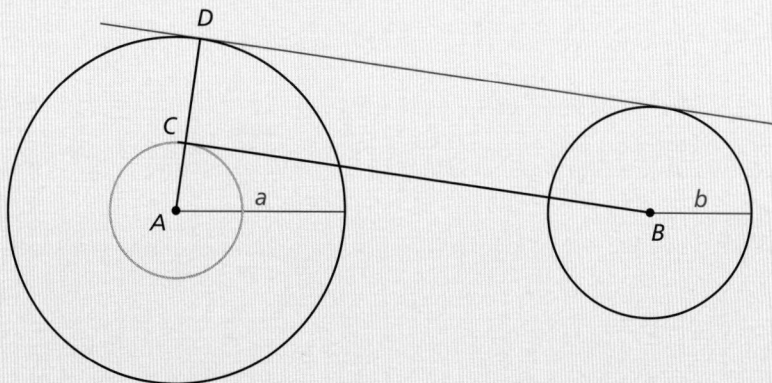

3 Draw two circles, and draw a common tangent by this method. Does it meet each circle only once?
4 Show why this method works.

5 Suppose two circles don't meet. How do you construct the common tangents which cross between the circles?

Pythagoras' theorem and circles

Many of the results of this chapter have been about right angles. Often we can use **Pythagoras' theorem** to find out distances concerning circles.

Examples   A point is 5 cm from the centre of a circle of radius 3 cm. What is the length of the tangent from the point to the circle?

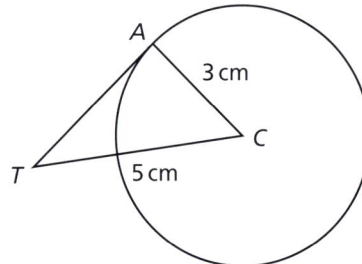

Remember:

Pythagoras' theorem – if c is the hypotenuse of a right-angled triangle, and a and b are the other sides, then $c^2 = a^2 + b^2$.

In the diagram, $TC = 5$ cm and $CA = 3$ cm. From theorem 1 on page 54, $\angle TAC = 90°$. TAC is a right-angled triangle with hypotenuse TC. Suppose $TA = x$ cm. From Pythagoras' theorem

$$x^2 + 3^2 = 5^2$$
$$x^2 = 25 - 9 = 16$$
$$x = 4$$

The length of the tangent is 4 cm.

A drain with diameter 20 cm is laid horizontally. There is water in it, to a depth of 3 cm. How wide is the water surface?

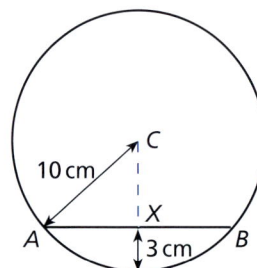

The radius of the pipe is 10 cm. The distance from the centre of the pipe to the top of the water is 10 cm − 3 cm, which is 7 cm.

The vertical line CX from the centre of the drain to the top of the water is at right angles to the water. Hence $\angle AXC = 90°$. Let $AX = x$ cm. Use Pythagoras' theorem in $\triangle AXC$.

$$x^2 + 7^2 = 10^2$$
$$x^2 = 100 - 49 = 51$$
$$x = \sqrt{51} = 7.14$$

The width of the water is twice this: 14.3 cm.

....

AB is the diameter of a circle, and C a point on the circle. $AC = 15$ cm and $BC = 17$ cm. Find the radius of the circle.

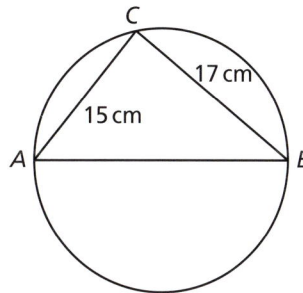

Let the diameter of the circle be x cm. We know that $\angle ACB = 90°$ (from theorem 4 on page 59). Using Pythagoras' theorem

$$x^2 = 15^2 + 17^2$$
$$= 225 + 289$$
$$= 514$$

So $x = \sqrt{514}$, which is 22.67. Halve this to get the radius.
The radius is 11.3 cm.

Exercise 5.8

1 A circle has radius 10 cm. Point T is 15 cm from the centre of the circle. How long is the tangent from T to the circle?

2 The tangent from X to a circle is 6 cm long. The distance from X to the centre of the circle is 7 cm. What is the radius of the circle?

3 The distance of a point P to the centre of a circle is 0.7 m. The radius of the circle is 0.3 m. Find the length of the tangent from P to the circle.

4 The tangent from a point to a circle is 1.4 m. The circle has radius 0.9 m. How far is the point from the centre of the circle?

5 From P to the centre of a circle is 18 cm, and the length of the tangent from P to the circle is 16 cm. What is the radius of the circle?

6 A circle has radius 88 mm. The tangent from a point to the circle is 109 mm long. Find the distance from the point to the centre of the circle.

7 A circle has radius 8 cm. A chord of the circle is 5 cm long. What is the distance from the centre of the circle to the centre of the chord?

8 A chord of a circle is 2 m long. The centre of the chord is 1.5 m from the centre of the circle. What is the radius of the circle?

9 A chord of a circle is 1.7 m long. The distance from the centre of the chord to the centre of the circle is 0.8 m. Find the radius of the circle.

10 The radius of a circle is 62 mm. The centre of the circle is 44 mm from the centre of a chord. Find the length of the chord.

11 A circle has radius 12 cm. A chord is 17 cm long. How far is it from the centre of the circle to the centre of the chord?

12 From the centre of a circle to the centre of a chord is 8 mm. The radius of the circle is 12 mm. Find the length of the chord.

13 *AB* is a diameter of a circle. *C* is a point on the circumference with *CA* = 3 cm and *CB* = 5 cm. Find the length of *AB*.

14 *AB* is a diameter of a circle, with *AB* = 1.4 m. *C* is a point on the circumference, with *CA* = 0.9 m. Find *CB*.

15 The distances from a point on a circle to the two ends of a diameter are 6 cm and 9 cm. Find the length of the diameter.

16 The radius of a circle is 8 cm. *X* is a point on the circumference, and the distance from *X* to one end of a diameter is 11 cm. Find the distance from *X* to the other end of the diameter.

17 A circle has radius 10 cm. Two chords *AB* and *CD* are parallel, with *AB* = 4 cm and *CD* = 5 cm. Find the distance between the chords if
   a the chords are on the same side of the centre of the circle
   b the chords are on opposite sides of the centre of the circle.

18 A square is drawn inside a circle of radius 2 m, with the corners on the circle. Find the side of the square.

19 A square of side 8 cm is drawn in a circle, with the corners on the circle. Find the radius of the circle.

20 A rectangle with sides 3 cm and 4 cm is drawn inside a circle, with the corners on the circle. Find the radius of the circle.

21 The diagram shows two wheels, of radii 5 cm and 8 cm, with their centres 16 cm apart. A belt passes tightly round both wheels. Find the length of each straight part of the belt.

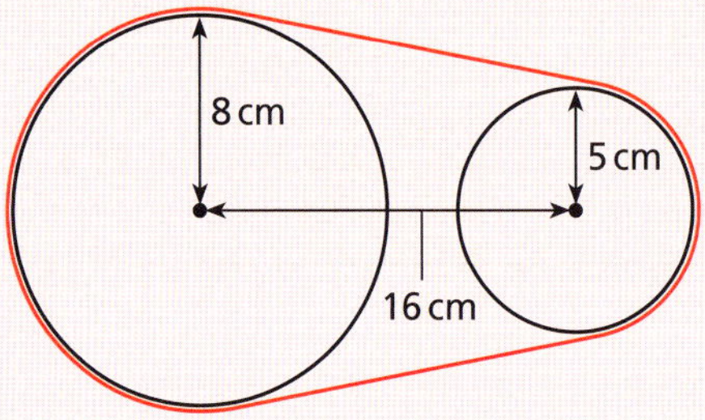

22 The belt of question 21 is rearranged so that it passes between the wheels as shown. Find the length of each straight part of the belt.

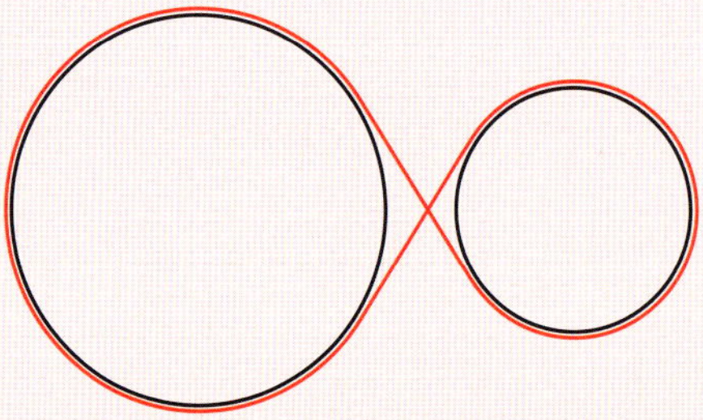

23 A tree trunk is a cylinder of width 20 cm. It is laid on the ground, and the top 3 cm is planed off. Find the width of the flat surface.

24 A tree trunk is a cylinder of diameter 25 cm. A beam of thickness 7 cm is to be cut from it. What is the maximum width of the beam?

25 A car wheel has radius 12 cm. A point on the ground is 30 cm from the base of the wheel. What is the distance from the point to the top of the wheel?

26 A circle with centre O has radius 6 cm. A point P is 2 cm from O. A chord AB goes through P, making 45° with OP. Find the length of AB.

27 An isosceles triangle with sides 5 cm, 5 cm and 6 cm is drawn in a circle, with corners on the circle.
   a Find the height of the triangle.
   b Let the radius of the circle be x cm. Find an equation in x.
   c Find the radius of the circle.

28 An isosceles triangle with sides 13 cm, 13 cm and 10 cm is drawn in a circle, with corners on the circle. Find the radius of the circle.

29 An equilateral triangle is drawn in a circle of radius 6 cm, with corners on the circle. Find the side of the triangle.

SUMMARY

■ A **tangent** touches a circle at one point only and is **perpendicular** to the radius at the point of contact.
■ The tangents to a circle from a point are equal in length.
■ The **perpendicular bisector** of a chord of a circle goes through the centre of the circle.
■ A diameter of a circle **subtends** a right angle at every point on the circumference.
■ These results combined with **Pythagoras' theorem** can be used to find distances concerning circles.

Exercise 5A

1 Make a copy of this diagram and draw the common tangents to the circles.

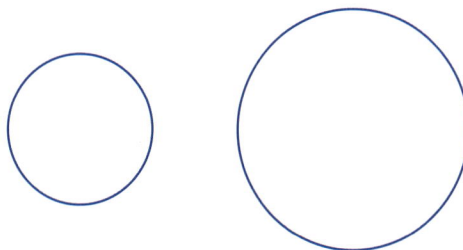

2 AB and CD are diameters of a circle. Show that the tangents at A, B, C and D form a parallelogram.

3 The tangents from T to a circle touch it at A and B. If $\angle ATB = 68°$ find $\angle TBA$.

4 A triangle has sides a, b and c. The radius of the *incircle* (the circle which touches all the sides of the triangle) is r. Show that the area of the triangle is $\frac{1}{2}r(a+b+c)$.

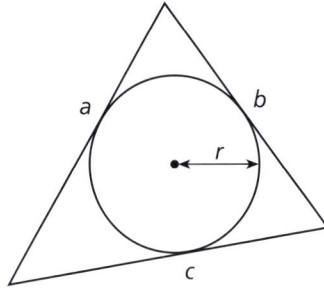

5 A circle has centre O. The midpoint of the chord AB is M. If $\angle AOM = 42°$ find $\angle MAO$.

6 AB is a diameter of a circle, and P is a point on its circumference. If $\angle PAB = 61°$ find $\angle PBA$.

7 A chord of a circle has length $7\,$cm. If the distance from the chord to the centre of the circle is $3\,$cm find the radius of the circle.

8 A point P is $8\,$cm from the centre of a circle. If the radius of the circle is $5\,$cm, find the length of the tangent from P to the circle.

9 XY is a diameter of a circle, and $XY = 0.8\,$m. Z is on the circumference, and $ZX = 0.6\,$m. Find ZY.

10 Two circles of radii $8\,$cm and $5\,$cm have their centres $10\,$cm apart. Find the length of a common tangent.

Exercise 5B

1 Two circles have radii $3\,$m and $4\,$m. Their centres are $8\,$m apart. How many common tangents are there to the circles?

2 The tangent from P to a circle meets the circle at A. The centre of the circle is O. If $\angle OPA = 37°$ find $\angle AOP$.

3 A and B are points on a circle. The tangents at A and B meet at T. If $\angle ABT = 51°$ find $\angle ATB$.

4 AB is a chord of a circle with centre O. The midpoint of AB is M. If $\angle OAB = 72°$ find $\angle MOA$.

5 PQ is a diameter of a circle, and R is a point on the circle. If $RQ = RP$ find $\angle PQR$.

6 AB is a diameter of a circle with centre O, and M is the midpoint of a chord AC. Show that MO is parallel to CB.

7 A plank AB leans against a roller as shown, touching it at X. The end B of the plank is $1.2\,$m from X, and the radius of the roller is $0.5\,$m. Find the distance to B from the centre of the roller.

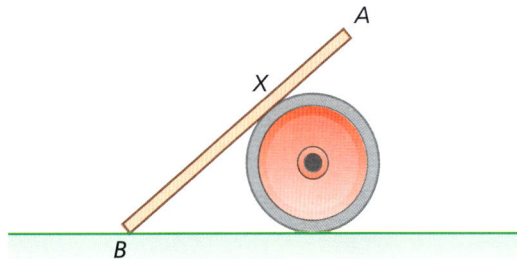

8 A chord of a circle is of length $6\,$cm, and the circle has radius $5\,$cm. Find the distance between the centre of the chord and the centre of the circle.

9 Two chords AB and AC are perpendicular. $AB = 0.3\,$m and $AC = 0.5\,$m. Find the radius of the circle.

10 A circle is inscribed in a regular hexagon of side $10\,$cm, that is the circle touches each side. Find the radius of the circle.

Exercise 5C (Ma1)

It has been known for over 2500 years that the Earth is round. How could we find its radius? It would be impossible to tie a tape measure around its circumference! This is one way to find the radius.

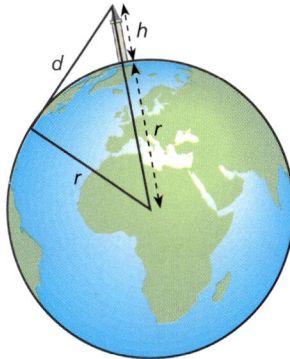

From the top of a tall tower, height h, a line to the horizon is a tangent to the Earth. Measure the distance d from the top of the tower to the horizon. Then if r is the radius of the Earth, Pythagoras' theorem gives

$$r^2 + d^2 = (r + h)^2$$

1 Expand and simplify this expression to get r in terms of d and h.
2 Suppose that the tower is 100 m high, and that the distance to the horizon is 36 km. What does this give for the radius of the Earth?
3 This is not an accurate way of measuring the radius. Can you explain why it is liable to error?
4 Another way of finding the radius, which was actually used in about 200 BC by Eratosthenes, involved measuring the angle of elevation of the Sun at two points on the Earth's surface. Describe an experiment to find the radius using this method.

Exercise 5D (Ma1)

1 Any triangle has an incircle, which touches all three sides. To construct the centre of the incircle, find the intersection of the bisectors of the angles. Draw a triangle and construct its incircle.
2 What about quadrilaterals? Not every quadrilateral has an incircle. Which of the quadrilaterals below have an incircle?

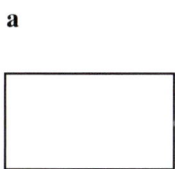

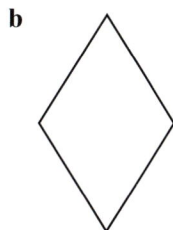

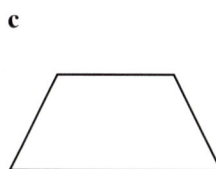

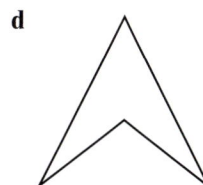

   a   **b**   **c**   **d**

3 Which of the following sorts of quadrilateral *must* have an incircle?

   square, rectangle, rhombus, kite, parallelogram, trapezium

4 Draw a circle and draw a quadrilateral round it, with each side on the circle. Measure the sides of the quadrilateral as accurately as you can. Can you find a relationship between the sides of the quadrilateral? Can you prove why it holds?

Exercise 5E ⬤Ma1

Circles seem to be such simple shapes, yet there is an astonishing number of theorems about them. In this chapter we have met only a few of them.

Many of the theorems and constructions concerning circles come from *The Elements* by Euclid, who lived about 300 BC. Below is his construction of a tangent from a point to a circle.

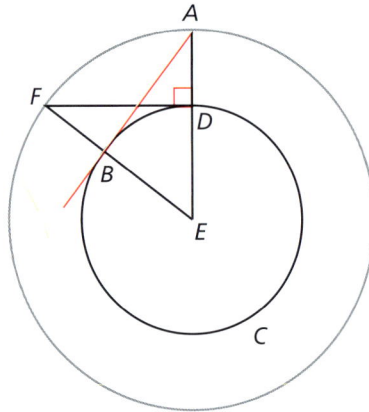

From a given point, to draw a straight line touching a given circle
Let A be the given point and BCD the given circle.
Let the centre of the circle be E: join AE. Draw the circle with centre E and radius EA.
From D draw DF at right angles to AE.
Join EF and AB. Then AB is the required tangent.

Try this construction. Is it more accurate than the method described in this chapter? Can you show why the construction works?

> Consider $\triangle AEB$ and $\triangle FED$.

6 Proportion

The simplest relationship between two variables occurs when one is a multiple of the other. If a car is being driven at constant speed, then the distance travelled is a multiple of the time taken. The mass of a substance is a multiple of the volume. In these cases, we say that one variable is **directly proportional** to the other. Suppose the variables are x and y. There is a special symbol, $\propto$, to give the relationship of proportionality.

$y \propto x$ means that y is proportional to x.

When y is proportional to x, it is a constant multiple of x. This constant multiple is the **constant of proportionality**.

If $y \propto x$, then $y = kx$ for some k.

> The expression $y \propto x$ is not an equation. It is converted into an equation, $y = kx$, by putting in the constant k.

If y and x are proportional, then the graph of y against x is a straight line through the origin, as shown. The gradient of the line is the constant of proportionality.

Note that $\dfrac{y}{x} = k$, which is constant. When two variables are proportional to each other, then their ratio is constant. Here are some examples of quantities which are proportional.

- Distance travelled is proportional to time. The constant of proportionality is the speed.
- Mass is proportional to volume. The constant of proportionality is the density.
- The electrical voltage across a circuit is proportional to the current. The constant of proportionality is the resistance.

You can find the constant of proportionality from a pair of values of the variables. Substitute the values, and you will have a simple equation in k.

Examples The variables S and T are proportional to each other. When $S = 20$ then $T = 25$. Find an equation giving T in terms of S.

We are told that $T \propto S$. Make this into an equation by putting in a constant of proportionality.

$$T = kS$$

To find k, put $S = 20$ and $T = 25$.

$$25 = 20k \qquad \text{hence } k = 1.25$$

The equation is $T = 1.25S$.

With S and T as above, find S when $T = 35$.

Put $T = 35$ into the equation.

$$35 = 1.25S$$

Hence $S = 35 \div 1.25 = 28$.
   The value of S is 28.

Exercise 6.1

1 The variables x and y are proportional. When $x = 4$ then $y = 10$.
  a Find an equation giving y in terms of x.
  b Find y when $x = 6$.
  c Find x when $y = 5$.

2 The variable m is proportional to the variable n. When $n = 0.2$ then $m = 1.6$.
  a Find an equation giving m in terms of n.
  b Find m when $n = 0.5$.
  c Find n when $m = 2$.

3 The variables p and q are proportional. When $q = 1\frac{1}{2}$ then $p = \frac{3}{8}$.
  a Find an equation giving q in terms of p.
  b Find q when $p = \frac{1}{3}$.
  c Find p when $q = 2\frac{1}{2}$.

4 The potential difference, V volts, across a circuit is proportional to the current, I amps, flowing through the circuit. When the current is 5 amps, the potential difference is 35 volts.
  a Find an equation giving V in terms of I.
  b Find the potential difference when the current is 7 amps.
  c Find the current when the potential difference is 14 volts.

5 *Hooke's law* states that the extension e m of a spring is proportional to the tension T newtons of the spring. When the tension is 30 newtons, the extension is 0.05 m.
  a Find e in terms of T.
  b Find the extension when the tension is 40 newtons.
  c Find the tension which causes an extension of 0.03 m.

6 If an amount of gas is kept at constant pressure, its volume V m^3 is proportional to its absolute temperature T K. Suppose the volume is 3 m^3 at 300 K.
  a Find V in terms of T.
  b What will the volume be at 450 K?
  c At what temperature will the volume be 0.5 m^3?

7 The force T N acting on a body is proportional to the acceleration a m/s^2 it produces (Newton's second law). Suppose a force of 10 N produces an acceleration of 0.2 m/s^2.
  a Find T in terms of a.
  b What force will accelerate the body at 0.5 m/s^2?
  c What acceleration is produced by a force of 70 N?

> Sir Isaac Newton (1642–1727), an English mathematician and scientist, laid the foundations of calculus and established the laws of motion and gravity.

8 The ratio $y : x$ is $4 : 5$. Write down an equation giving y in terms of x.

9 The ratio $p : q$ is $\frac{1}{2} : \frac{3}{4}$. Write down an equation giving p in terms of q.

Inverse proportion

When two variables are directly proportional, they increase together and decrease together. In some other situations, when one variable increases the other decreases. For example, the greater the speed of a car the less time it will take over a journey. In many cases like this we say that one variable is **inversely proportional** to the other. This is also written using the $\propto$ symbol.

$$y \propto \frac{1}{x} \qquad y \text{ is inversely proportional to } x$$

The statement of inverse proportionality can be converted to an equation by introducing a constant of proportionality.

$$\text{If } y \propto \frac{1}{x}, \text{ then } y = \frac{k}{x}.$$

If y and x are inversely proportional, then the graph of y against x has the shape shown.

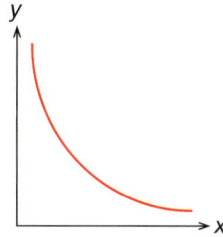

Note that $k = xy$. If x and y are inversely proportional, then their product is constant.

Boyle's law on the relationship between the pressure P and the volume V of a gas can be phrased in two ways.

$$\text{The volume is inversely proportional to the pressure: } V \propto \frac{1}{P}$$

$$\text{The product of the pressure and the volume is constant: } PV = \text{constant}$$

Example   Suppose m and n are inversely proportional, and that $m = 3$ when $n = 4$.

a Find an equation giving m in terms of n.
b Find the value of m when $n = 24$.
c Find the value of n when $m = \frac{1}{4}$.

a We are told that $m \propto \frac{1}{n}$. This is equivalent to the equation $m = \frac{k}{n}$. Put in the values given.

$$3 = \frac{k}{4} \quad \text{hence } 3 \times 4 = k$$

The equation is $m = \frac{12}{n}$.

b Put $n = 24$.

$$m = \tfrac{12}{24} = \tfrac{1}{2}$$

When $n = 24$, then $m = \tfrac{1}{2}$.

c Put $m = \tfrac{1}{4}$.

$$\tfrac{1}{4} = \frac{12}{n}$$

$$n = 12 \div \tfrac{1}{4} = 12 \times 4 = 48$$

When $m = \tfrac{1}{4}$, then $n = 48$.

Exercise 6.2

1 y is inversely proportional to x. When $x = 3$, then $y = 15$.
   a Find an equation giving y in terms of x.
   b Find y when $x = 12$.
   c Find x when $y = 3$.

2 The quantities p and q are inversely proportional. When $q = 60$, then $p = 0.4$.
   a Find an equation giving p in terms of q.
   b Find q when $p = 0.1$.
   c Find p when $q = 100$.

3 *Boyle's law* states that if a mass of gas is kept at constant temperature, its volume $V\,\text{m}^3$ is inversely proportional to its pressure P pascals. At a pressure of 200 pascals, an amount of gas had volume $1.8\,\text{m}^3$.
   a Find an equation giving V in terms of P.
   b If the pressure is reduced to 160 pascals, what would the volume be?
   c What pressure would reduce the volume to $0.6\,\text{m}^3$?

4 The electric current, I amperes, in a wire is inversely proportional to the resistance, R ohms. When the resistance is 100 ohms, the current is 2.5 amps.
   a Find an equation giving I in terms of R.
   b Find the current when the resistance is 80 ohms.
   c Find the resistance when the current is 0.1 amps.

5 The quantities M and N are inversely proportional. When $M = 3$, then $N = 0.4$.
   a Find an equation giving M in terms of N.
   b Find M when $N = 1.2$.
   c For what value of M are M and N equal?

6 The electrical resistance, R ohms, of a fixed length of wire is inversely proportional to its cross-sectional area, $a\,\text{mm}^2$. For wire of cross-sectional area $0.05\,\text{mm}^2$ the resistance is 10 ohms.
   a Find an equation giving R in terms of a.
   b If the cross-section is a circle of radius 0.2 mm, find the resistance of the wire.
   c If the wire has circular cross-section, find the radius if the resistance is 20 ohms.

7 The product of two quantities x and y is constant. Write down a proportionality statement between x and y.

Proportionality to powers

In many cases, one variable is proportional to a power of another variable. It might be that y is proportional to the square of x, or that m is proportional to the fourth power of n. These are again written using the $\propto$ symbol.

$$y \propto x^2 \qquad m \propto n^4$$

The statement of proportionality can be converted to an equation by a constant of proportionality. If you know a pair of values of the variables, then proceed exactly as before. For example, the area A of a circle is proportional to the square of the radius r.

$$A \propto r^2$$

The constant of proportionality is of course π.

$$A = \pi r^2$$

If y is proportional to x^2, then the graph of y against x has the shape shown.

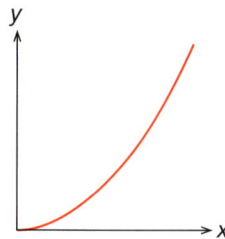

Example M is proportional to the square of N. When $N = 2$, then $M = 100$.

   a   Find an equation giving M in terms of N.
   b   Find the positive value of N when $M = 36$.

   a   We are given that $M \propto N^2$. Write this as the equation $M = kN^2$. Put in the values of M and N.

$$100 = k \times 2^2 \quad \text{hence } k = 100 \div 4 = 25$$

    The equation is $M = 25N^2$.

   b   Put in $M = 36$.

$$36 = 25N^2$$
$$N^2 = 1.44 \quad \text{hence } N = 1.2$$

    When $M = 36$, then $N = 1.2$.

Exercise 6.3

1 y is proportional to the square of x. When $x = 3$, then $y = 1.35$.
  a Find an equation giving y in terms of x.
  b Find y when $x = 1.2$.
  c Find x when $y = 21.6$.
2 P is proportional to the cube of Q. When $Q = 5$, then $P = 0.02$.
  a Find P in terms of Q.
  b Find P when $Q = 4$.
  c Find Q when $P = 0.0025$.
3 M is proportional to the square of N. When $N = \frac{1}{4}$, then $M = 3$.
  a Find an equation giving M in terms of N.
  b Find M when $N = 1\frac{1}{2}$.
  c Find N when $M = \frac{1}{3}$.
4 Spheres of different radius are made out of the same metal. The mass m g of the sphere is proportional to the cube of the radius r cm. A sphere of radius 2 cm has mass 220 grams.
  a Find m in terms of r.
  b Find the mass of a sphere with radius 5 cm.
  c Find the radius of a sphere with mass 100 g.
5 The time of swing, s seconds, of a pendulum is proportional to the square root of its length, l metres. A pendulum 2 m long has time of swing 2.84 seconds.
  a Find s in terms of l.
  b If a pendulum is 0.5 m long, find its time of swing.
  c What length of pendulum would have a time of swing of exactly 1 second?
6 When a body is moving rapidly through the air, the air resistance, R newtons, is proportional to the square of the velocity, v m/s. At a velocity of 100 m/s the air resistance is 50 newtons.
  a Find R in terms of v.
  b Find the resistance at 150 m/s.
  c Find the velocity which gives a resistance of 80 newtons.
7 z is proportional to the square of y, and y is proportional to the cube of x. What is the proportionality relationship between z and x?
8 k is proportional to the square of m, and to the cube of n. What is the proportionality relationship between m and n?
9 The ratio of y to x^2 is $3:4$. Find a proportionality statement between y and x. Find an equation giving y in terms of x.
10 The ratio of P to Q^3 is $\frac{1}{2}:4$. Find a proportionality statement between P and Q. Find an equation giving P in terms of Q.

In some situations, one variable is inversely proportional to a power of another. A famous case is *Newton's law of gravity*.

> The attraction between two bodies is inversely proportional to the square of the distance between them.

These cases are handled in the same way as other cases of proportion.

Example y is inversely proportional to the cube of x. When $x = 4$, then $y = 25$.

a Find an equation giving y in terms of x.
b Find x when $y = 0.2$.

a We are given that $y \propto \dfrac{1}{x^3}$. Write this as $y = \dfrac{k}{x^3}$ and put in the values.

$$25 = \frac{k}{4^3} \quad \text{hence } k = 25 \times 64 = 1600$$

The equation is $y = \dfrac{1600}{x^3}$.

b Now put $y = 0.2$.

$$0.2 = \frac{1600}{x^3} \quad \text{hence } x^3 = 1600 \div 0.2 = 8000$$

So $x^3 = 8000$, giving $x = 20$.
When $y = 0.2$, then $x = 20$.

Exercise 6.4

1 p is inversely proportional to the square of q. When $p = 6$, then $q = 1.25$.
  a Find an equation giving p in terms of q.
  b Find p when $q = 8$.
  c Find the positive value of q when $p = 150$.
2 W is inversely proportional to the cube of T. When $T = 2$, then $W = 80$.
  a Find an equation giving W in terms of T.
  b Find W when $T = 4$.
  c Find T when $W = 40$.
3 P is inversely proportional to the square root of Q. When $Q = 8$, then $P = 3$.
  a Find an equation giving P in terms of Q.
  b Find P when $Q = 18$.
  c Find Q when $P = \frac{1}{2}$.
4 The illumination from a light is inversely proportional to the square of the distance from the light. At a distance of 6 m the illumination is 10 candelas. What is the illumination at a distance of 10 m?
5 Suppose that two masses, 1 m apart, exert a force on each other of 2.3×10^{-11} newtons. What would be the force if they were 2.5 m apart? At what distance apart would the force be 1.7×10^{-11} newtons?

> **Remember:**
>
> *Newton's law of gravity – The attraction between two bodies is inversely proportional to the square of the distance between them.*

6 The electrical resistance, R ohms, of a fixed length of wire is inversely proportional to the square of its diameter, d mm. A length of wire of diameter 0.4 mm has resistance 20 ohms.
  a Find an equation giving R in terms of d.
  b What is the resistance of a wire of diameter 0.1 mm?
  c What diameter of wire would give a resistance of 5 ohms?
7 The product of y and x^2 is constant. Write down a proportionality relationship between x and y.
8 The product of t and $\sqrt{l}$ is constant. Write down a proportionality relationship between l and t.

Finding the power

Suppose you think that a quantity is proportional to a power of another, but you don't know the actual power. You can find the power and the constant of proportionality from *two* pairs of values of the quantities.

Example

We know that y is proportional to a power of x. We have found that $y = 0.3$ when $x = 4$, and that $y = 2.4$ when $x = 8$. Find the equation giving y in terms of x.

Put $y \propto x^n$, where n is the unknown power. This gives the equation $y = kx^n$, where k is the constant of proportionality. Put in the known values

$$0.3 = k4^n \quad \text{and} \quad 2.4 = k8^n$$

Divide the second equation by the first, to get rid of the k.

Remember chapter 1
$$\left(\frac{a}{b}\right)^n = \frac{a^n}{b^n}$$

$$\frac{2.4}{0.3} = \frac{8^n}{4^n}$$

$$\frac{2.4}{0.3} = 8 \quad \text{and} \quad \frac{8^n}{4^n} = \left(\frac{8}{4}\right)^n = 2^n$$

Hence $8 = 2^n$. This gives $n = 3$.
   We now know that $0.3 = k4^3 = 64k$. This gives $k = 0.3 \div 64$, that is, $k = 0.004\,687\,5$.
   The relationship is $y = 0.004\,687\,5x^3$.

Exercise 6.5

1 We know that m is proportional to a power of n. We have found that $m = 5$ when $n = 2$, and $m = 20$ when $n = 4$. Find the equation giving m in terms of n.

2 We know that y is proportional to a power of x. We find that $y = 120$ when $x = 3$, and $y = 7.5$ when $x = 1.5$. Find the equation giving y in terms of x.

3 We know that P is proportional to a power of Q. When $Q = 12$ then $P = 14$, and when $Q = 3$ then $P = 7$. Find an equation giving P in terms of Q.

4 We know that F is inversely proportional to a power of G. When $G = 10$ then $F = 8$, and when $G = 20$ then $F = 2$. Find the equation giving F in terms of G.

5 We know that y is inversely proportional to a power of x. When $x = 2$ then $y = 48$, and when $x = 4$ then $y = 6$. Find the equation giving y in terms of x.

6 Suppose $y \propto x^2$. What power of y is x proportional to?

7 Suppose $p \propto \dfrac{1}{q^n}$. What power of p is q proportional to?

8 Suppose the volume of a cube is V and its surface area is A. Write down a proportionality statement for V in terms of A.

SUMMARY

■ If two variables increase and decrease at the same rate, so that for example if one is doubled the other is also doubled, they are **directly proportional** to each other. In this case the ratio between the variables is constant. This can be written using the $\propto$ symbol. The statement of proportionality can be converted to an equation by inserting a **constant of proportionality**. Write $y \propto x$, or $y = kx$.

■ If two variables increase and decrease at opposite rates, then they are **inversely proportional** to each other. In this case the product of the variables is constant. Write $y \propto \dfrac{1}{x}$, or $y = \dfrac{k}{x}$.

■ In many situations one variable is proportional or inversely proportional to a power of another variable. For example, $y \propto x^2$, or $y \propto \dfrac{1}{x^3}$.

Exercise 6A

1 y is proportional to x. When $x = 0.7$, then $y = 280$. Find an equation giving y in terms of x.

2 p is inversely proportional to q. When $p = 20$, then $q = 1.2$. Find an equation giving p in terms of q.

3 At a height of h m above sea level, the distance d m to the horizon is proportional to the square root of h. From the top of a 20 m mast, the horizon is 11 000 m away.
   a Find an equation giving d in terms of h.
   b How far can you see from the top of a 30 m mast?
   c How high must you be to see for 20 000 m?

4 Suppose the string of a musical instrument is held at constant tension and length. Then the frequency, f cycles per second, is inversely proportional to the square root of its density, d grams per metre (g/m). A string of density 0.2 g/m has frequency 260 cycles per second (middle C).
   a Find an equation giving f in terms of d.
   b What is the frequency if the density is changed to 0.3 g/m?
   c What density is needed for a note of G (390 cycles per second)?

5 Suppose that y is proportional to a power of x. When $x = 100$ then $y = 4$, and when $x = 300$ then $y = 108$. Find an equation giving y in terms of x.

6 Suppose u is inversely proportional to the cube root of v. What is the proportionality statement giving v in terms of u?

Exercise 6B

1 F is proportional to G. When $G = 25$, $F = 40$. Find an equation giving F in terms of G.

2 r is inversely proportional to s, and $rs = 60$. Find an equation giving r in terms of s.

3 If the string of a musical instrument has fixed length and density, then its frequency, f cycles per second, is proportional to the square root of the tension, T newtons. Suppose the string playing middle C (260 cycles/s) has tension 200 newtons.
   a Find an equation giving f in terms of T.
   b If the tension is increased to 300 newtons, what is the frequency?
   c What should the tension be for a note of F (347 cycles/s)?

4 Suppose a steel sphere is free to rotate about a diameter. If a constant torque is applied to the sphere for 1 second, the speed, s revolutions per second, is inversely proportional to the fifth power of its radius, r cm. Suppose that a sphere of radius 10 cm reaches a speed of 5 revolutions per second.
   a Find an equation giving s in terms of r.
   b Find the speed of a sphere of radius 7 cm.
   c For what radius would the sphere reach a speed of 100 revolutions per second?

5 Suppose y is inversely proportional to a power of x. When $x = 8$ then $y = 54$, and when $x = 72$ then $y = 18$. Find an equation giving y in terms of x.

6 y is proportional to the cube of x, and x is inversely proportional to the square root of z. What is the proportionality relationship between y and z?

Exercise 6C Ma1

Johannes Kepler (1571–1630) formulated three laws of planetary motion which were vital in the development of astronomy. He was more proud, however, of his model of the universe which fitted the orbits of the planets within regular solids (octahedron, icosahedron, dodecahedron, tetrahedron and cube).

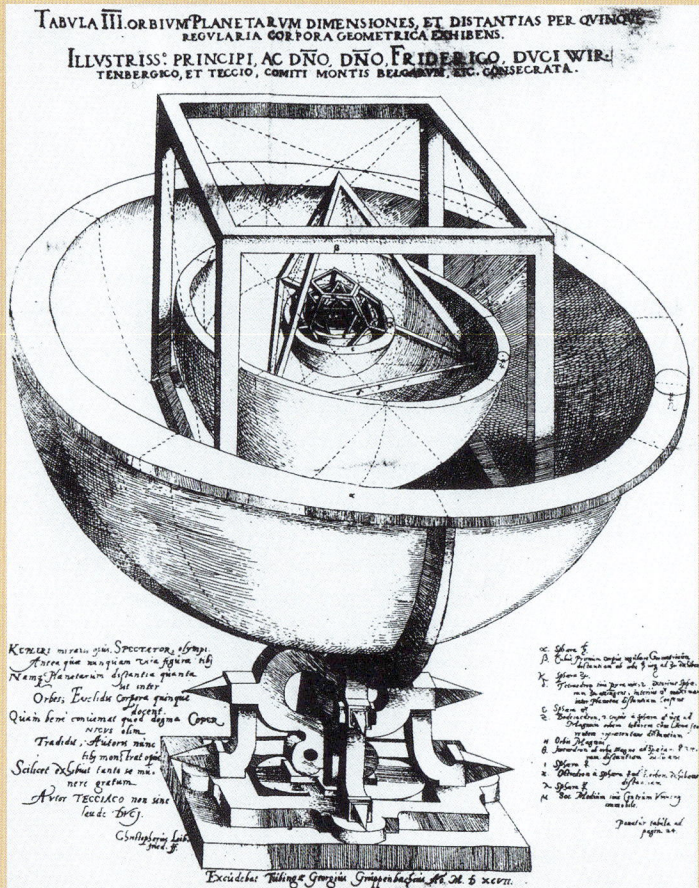

A famous example of proportionality is *Kepler's third law*. It relates the distance to a planet from the Sun with the length of its year.

Suppose a planet is d km from the Sun, and its year is y (Earth) days. The law states that the cube of the distance is proportional to the square of the year. So $d^3 \propto y^2$.

1 The Earth is about 150 000 000 km from the Sun. Find the constant of proportionality.
2 Venus is 108 000 000 km from the Sun, and the Venusian year is 225 days. Check that Venus obeys Kepler's law.
3 Mars is 228 000 000 km from the Sun. How long is the Martian year?
4 Jupiter has a year of length 4330 days. How far is it from the Sun?

Exercise 6D 〔Ma1〕

Equations of proportionality involve only two variables. Most things are more complicated than that! The volume of a cylinder, for example, involves both the height and the radius. The formula is $V = \pi r^2 h$. We say V is jointly proportional to h and the square of r.

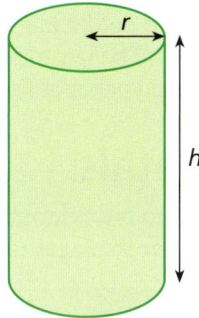

1 Suppose r is doubled and h is tripled. What happens to the volume?
2 Suppose r is doubled and h halved. What happens to the volume?
3 Suppose r is doubled. What would we have to divide h by if the volume is to remain the same?
4 Suppose h is doubled. What would we have to divide r by if the volume is to remain the same?

The pressure of a fixed mass of gas is given by $P = \dfrac{kT}{V}$, where k is a constant, T is the absolute temperature and V is the volume. So P is jointly proportional to T and inversely proportional to V.

5 Suppose T is doubled and V is tripled. What happens to the pressure?
6 Suppose V is doubled. How must T change if the pressure is to remain the same?

The following exercise extends the material of this chapter to a topic beyond GCSE.

Exercise 6E 〔Ma1〕

The last section of this chapter asked you to find the power of a proportionality relationship. If the power is a square or cube, then it can be spotted fairly easily. But if the power is not a whole number then it is more difficult.

Suppose that $y = kx^n$, that is, y is proportional to the nth power of x. On your calculator you will find a button labelled log.
If you apply the log function to the equation above, it becomes

> Logarithms were defined in exercise 1E, and used in exercise 2E.

$$\log y = \log k + n \log x$$

So by taking logs we can reduce the proportionality statement to a linear equation.

1 Suppose V volts are applied to an electric lamp. The luminosity, C candelas, is proportional to a power of V, that is, $C = kV^n$ for some n. Suppose that $C = 21$ for $V = 10$, and that $C = 240$ for $V = 18$.
  Use your calculator to find the logs of these values, and hence find two simultaneous equations in $\log k$ and n. Eliminate $\log k$ and find n. Find the equation giving C in terms of V.
2 Suppose an insulator is d mm thick. If it will break down at a voltage of V volts, then V is proportional to a power of d. When $d = 1$ then $V = 100000$, and when $d = 4$ then $V = 280000$. Find an equation giving V in terms of d.

7 Cumulative frequency

Vessa got 56% in the first paper of her GCSE Maths exam. Unfortunately she was ill for the second paper, and had to miss it. What should the examiners give her as a mark for the second paper?

It might not be fair to give her 56% for the second paper. If the second paper were much harder than the first, then she probably would not have got as high a mark as 56%. It would not be fair to give the same mark for the second paper as for the first, but it would be fair to put her in the same *order* (with other students in the class) for the second paper as for the first. We will return to this problem later.

Running totals

Tarquin is saving up to buy a mountain bike which costs £100. He keeps a record of how much he manages to save each month.

month	January	February	March	April	May
amount (£)	17	35	25	11	20

Of course, what he wants to know is the total he has saved at the end of each month. To find this, add up the amounts saved. At the end of February, for example, he will have saved £17 + £35, which is £52. These totals are called **running totals**.

month	January	February	March	April	May
amount (£)	17	35	25	11	20
running total (£)	17	52	77	88	108

By the end of May he has over £100, so he can buy the bike.

Exercise 7.1

1 A church is appealing for funds to restore its steeple. The table below gives the amounts received over six months. Write out an extra row showing the running total of the amount received.

month	March	April	May	June	July	August
amount (£)	243	433	549	430	395	218

2 A motorist keeps a record of the number of miles he has driven each week. The table below gives the results. Write out an extra row showing the running total of the distance he has driven.

week	1	2	3	4	5
distance (miles)	173	214	153	203	188

3 The monthly sales of a textbook are shown below. Write out an extra row showing the running total of the sales.

month	August	September	October	November	December
sales	684	2404	1855	451	193

Cumulative frequency tables

You have already met frequency tables. The table below gives the midday temperatures over a period of 200 days.

temperature (°F)	50–	60–	70–	80–	90–100
frequency	17	46	73	52	12

From this table you can read off, for example, the number of days in which the temperature was in the eighties.

You might also want to know the number of days in which the temperature was below 80 °F, or the number of days in which the temperature was below 70 °F. This is easily done. Construct an extra row, giving the running total of the frequencies. The running total is the **cumulative frequency**.

temperature (°F)	50–	60–	70–	80–	90–100
frequency	17	46	73	52	12
cumulative frequency	17	63	136	188	200

So we can read off from the table that 136 days had a temperature below 80 °F.

● *In general, the cumulative frequency at a value gives the number of data less than the value. It is found by calculating a running total of all the frequencies up to the value.*

Note. You always have a check on your arithmetic. The cumulative frequency at the end must be the total number of data. In the table above, the cumulative frequency at the end is 200.

Exercise 7.2

1 The frequency table below gives the marks obtained in an exam by 200 candidates. Write out an extra row for the cumulative frequencies.

mark	0–19	20–39	40–59	60–79	80–100
frequency	7	24	83	52	34

   a How many candidates got less than 40?
   b How many candidates got at least 60?

2 The prices of houses for sale in a town were found. The frequency table below gives the results. Write out an extra row for the cumulative frequencies.

price (£1000s)	50–	60–	70–	80–	90–100
frequency	23	48	107	90	32

   a How many houses cost less than £80000?
   b How many houses cost at least £60000?

3 The ages of 120 people at a wedding are given in the frequency table below. Write out an extra row for the cumulative frequencies.

age	0–	10–	20–	30–	40–	50–	60–70
frequency	5	9	35	28	21	17	5

   a How many people were at least 20 years old?
   b How many people were under 50 years old?

4 The times taken by 80 people to run a race are given in the frequency table below. Write out an extra row for the cumulative frequencies.

time (seconds)	150–	155–	160–	165–	170–175
frequency	35	18	12	9	6

   a How many people took under 160 seconds?
   b How many people took at least 155 seconds?

Cumulative frequency curve

Not many people can look at a table of figures and draw useful conclusions! Pictures are always helpful. For cumulative frequency, the picture is a **cumulative frequency curve** or **ogive**. With this curve, you can just read off the cumulative frequency at any value.

Consider again the temperature example on page 81. Here is the table giving the temperature intervals, the frequencies and the cumulative frequencies.

temperature (°F)	50–	60–	70–	80–	90–100
frequency	17	46	73	52	12
cumulative frequency	17	63	136	188	200

Put the quantity we are measuring along the x-axis, and the cumulative frequency up the y-axis. So the temperatures go along the x-axis, and the cumulative frequencies up the y-axis. The lowest temperature was 50 °F, so start the x-axis at 50.

The cumulative frequency at 60 °F was 17. This means that the temperature was less than 60 °F on 17 days. So we plot a point at (60, 17). Note that this point is the upper limit of the interval 50– as all the values in the interval are less than 60. Here are all the values to plot.

0 days had a temperature under 50 °F.	Plot a point at (50, 0).
17 days had a temperature under 60 °F.	Plot a point at (60, 17).
63 days had a temperature under 70 °F.	Plot a point at (70, 63).
136 days had a temperature under 80 °F.	Plot a point at (80, 136).
188 days had a temperature under 90 °F.	Plot a point at (90, 188).
200 days had a temperature under 100 °F.	Plot a point at (100, 200).

So we plot points at (50, 0), (60, 17), (70, 63), (80, 136), (90, 188) and (100, 200). Now join them up by a smooth curve. The final result is shown in the diagram.

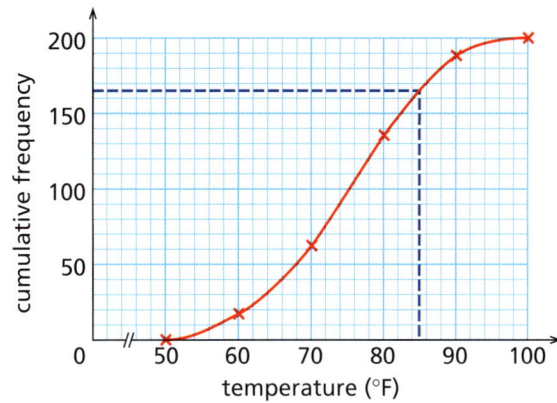

Example From the cumulative frequency curve above, find out on how many days a temperature of less than 85 °F was recorded.

Take 85 on the *x*-axis, go up until you meet the curve, then go across to the *y*-axis. This is shown by a dark blue dotted line on the diagram. A temperature of 85 °F corresponds to a cumulative frequency of 165.
  The temperature was less than 85 °F on 165 days.

Exercise 7.3

1 The cumulative frequency graph gives the weights of 100 cats. How many cats were under 4.5 kg?

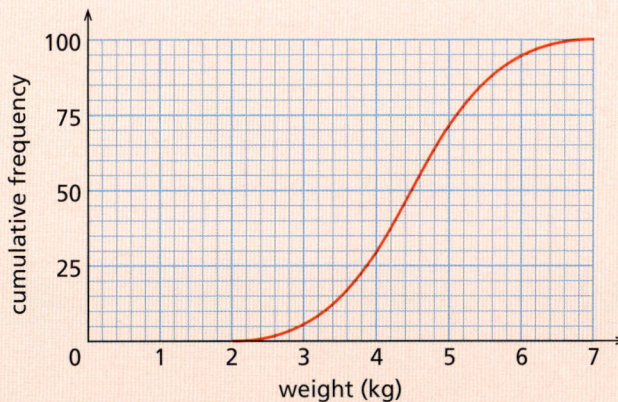

2 The cumulative frequency graph gives the distances travelled in a week by 40 motorists. How many travelled less than 250 miles?

3 The prices of 100 second-hand cars advertised in a newspaper were found, and the information displayed on a cumulative frequency graph as shown.
   a How many cars cost less than £4000?
   b How many cars cost at least £2000?

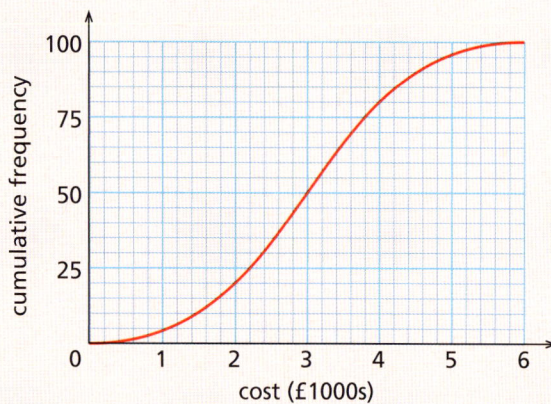

4 The cumulative frequency graph gives the distance travelled to school by 80 students.
   a How many students travelled less than 12 miles?
   b How many students travelled at least 15 miles?

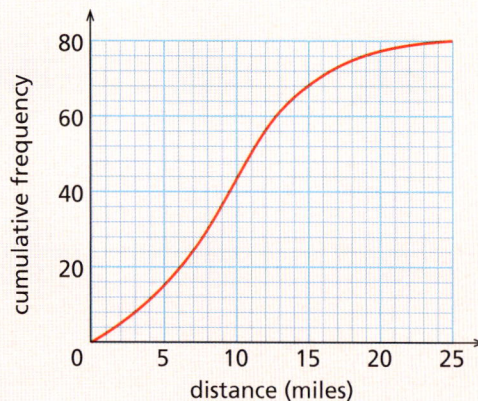

Questions 5–8 ask you to draw graphs using the data sets of exercise 7.2. You will be using these graphs later on in this chapter, so draw them neatly and keep them.

You will need:
- ruler
- sharp pencil
- graph paper

5 Question 1 of exercise 7.2 gave the marks obtained in an exam by 200 candidates. Construct a cumulative frequency graph on a diagram similar to the one shown here.
   a How many candidates got under 95 marks?
   b How many candidates got at least 75 marks?

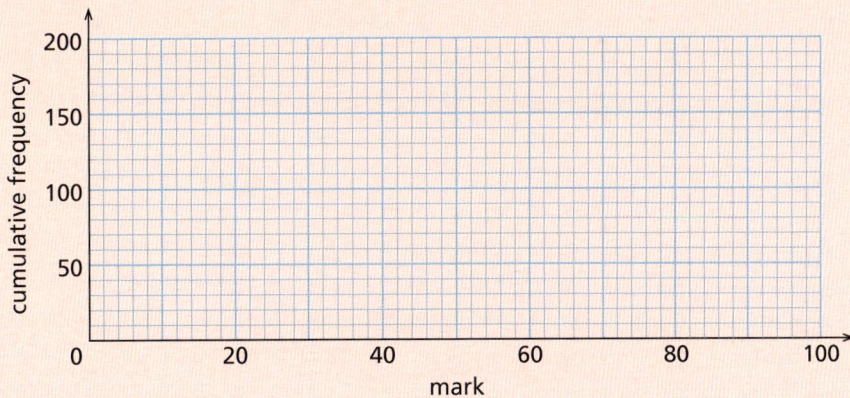

6 Question 2 of exercise 7.2 gave the prices of houses in a town. Construct a cumulative frequency graph on a diagram similar to the one shown.
   a How many houses cost less than £85 000?
   b How many houses cost at least £66 000?

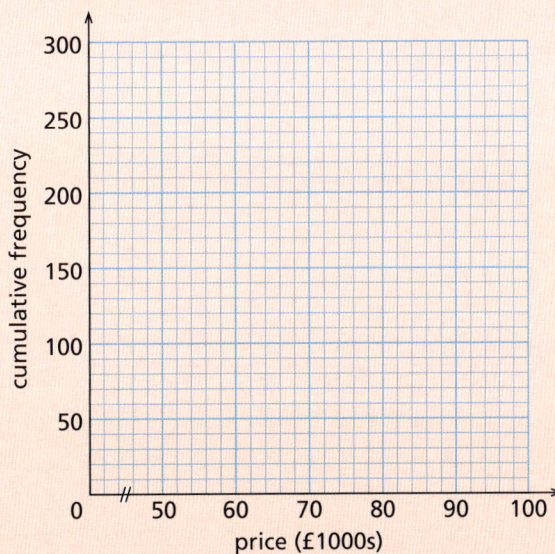

7 Question 3 of exercise 7.2 gave the ages of 120 people at a wedding. Construct a cumulative frequency graph on a diagram similar to the one shown.
   a How many people were aged under 45?
   b How many people were aged between 25 and 45?

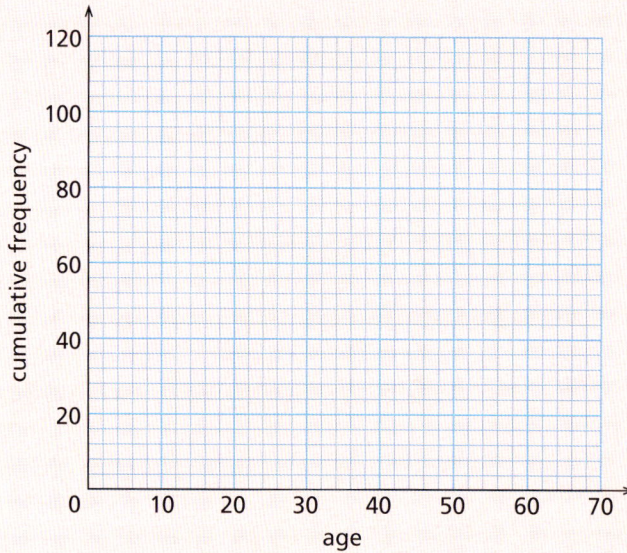

8 Question 4 of exercise 7.2 gave the race times of 80 runners. Construct a cumulative frequency graph on a diagram similar to the graph shown.
   a How many people took less than 163 seconds?
   b How many people took between 152 seconds and 163 seconds?

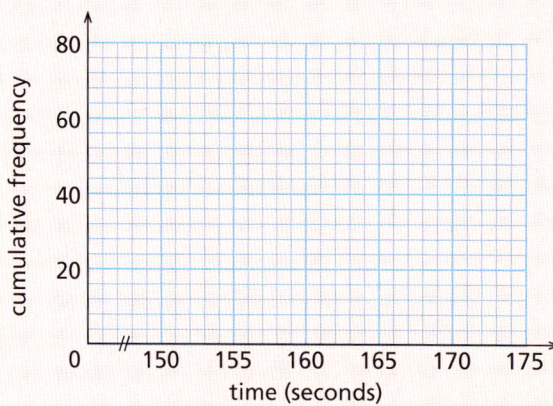

Median and quartiles

Recall that the **median** of a set of data is the middle value. You have already found the median of small data sets – you arrange them in order and pick the middle value.

This isn't possible when you are given data in a frequency table – for a start, you don't know what the values actually were, just how many of each different sort there were. Also there may be too many values; it would take a long time to sort 200 temperatures in order. So, instead, we use the cumulative frequency curve.

What is the cumulative frequency at the median? Look at the cumulative frequency graph for the temperatures (page 83). There are 200 temperatures, and we want to separate out the top half from the bottom half. So, at the median, the cumulative frequency is 100.

In general the median is the middle number, which cuts off the top half from the bottom half. So half the data is less than the median and half is greater. At the median, the cumulative frequency is always half the total frequency. So the median is at the point half-way up the graph.

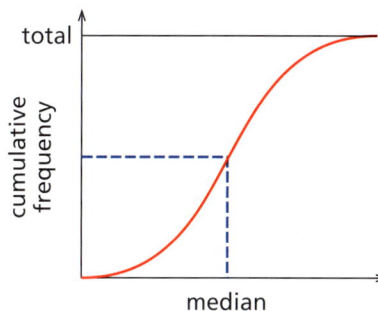

> **Note.** The median is not half the total frequency. It is the value which corresponds to half the total frequency.

Example Find the median temperature for the example above.

There are 200 days. Half of this is 100. The median temperature will have a cumulative frequency of 100.

Find 100 on the *y*-axis, go across to the graph, and down to the *x*-axis. This is shown as a blue dotted line. You have found the temperature, 76 °F, which has cumulative frequency 100. This temperature is the median.

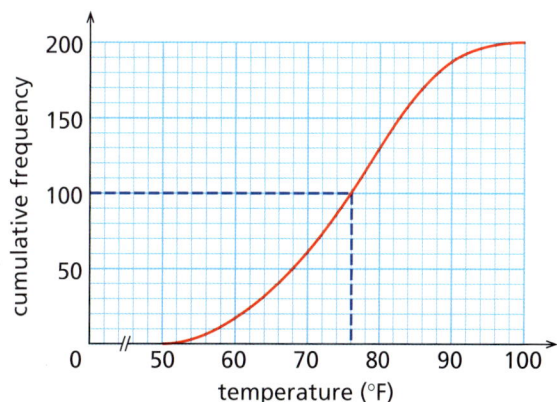

The median temperature is 76 °F.

Note. Strictly, the median should occur between the 100th and the 101th temperatures, i.e. the 100.5th temperature. But it is impossible to pick out 100.5 on the *y*-axis!

Exercise 7.4

In each of these questions, use the cumulative frequency graph for the corresponding question in exercise 7.3 to find:

1 the median weight of the cats
2 the median distance travelled
3 the median price of the cars
4 the median distance travelled
5 the median mark
6 the median house price
7 the median age
8 the median race time.

Quartiles

> The lower quartile, median and upper quartile are sometimes called Q_1, Q_2 and Q_3 respectively.

The median is the value which cuts off the top half of the data from the bottom half. You might also want to find the top *quarter* of the data, or the bottom quarter; for example, the top quarter of the candidates for an exam might be awarded a distinction. The values which cut off the top and bottom quarters are the **quartiles**.

The **upper quartile** is the value which cuts off the top quarter of the data from the bottom three quarters.
The **lower quartile** is the value which cuts off the bottom quarter of the data from the top three quarters.

You can find the quartiles from the cumulative frequency curve. Find a quarter of the total frequency. Go to this point on the y-axis. Go across to the curve, and down to the x-axis. This gives the lower quartile. The upper quartile is found similarly.

So the quartiles are at a quarter and at three quarters of the total height of the graph.

Example Find the quartiles of the data in the temperature example on pages 82–83.

There are 200 days. A quarter of 200 is 50. Find 50 on the y-axis, go across to the curve, and down to the x-axis, at 68. The line is shown in green.

Three quarters of 200 is 150. Find 150 on the y-axis, go across to the curve, and down to the x-axis, at 83. The line is shown in dark blue.

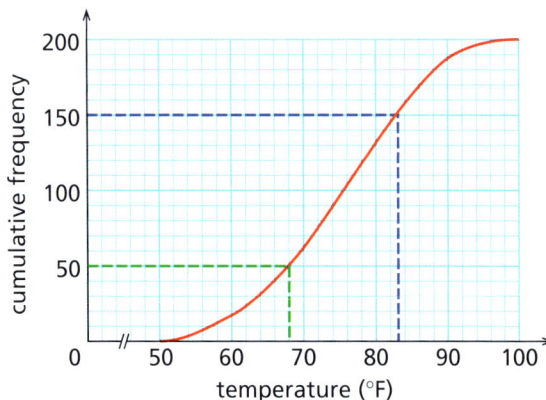

The lower quartile is 68 °F, and the upper quartile is 83 °F.

Exercise 7.5

In each of questions 1–8, use the cumulative frequency graph for the corresponding question in exercise 7.3 to find:

1 the quartile weights of the cats
2 the quartile distances travelled
3 the quartile prices of the cars
4 the quartile distances travelled
5 the quartile marks
6 the quartile house prices
7 the quartile ages
8 the quartile race times.
9 A golfer goes round a course 60 times. The frequency table below gives the number of strokes he took.

number of strokes	80–84	85–89	90–94	95–99	100–104
frequency	4	13	17	15	11

  a Find the cumulative frequencies.
  b Plot a cumulative frequency graph.
  c Find the median number of strokes.
  d Find the upper and lower quartile numbers of strokes.
10 A survey was made of the mean amount spent by the customers in 100 pizza restaurants. The results are shown below.

mean amount spent (£)	3–	4–	5–	6–	7–	8–9
frequency	8	16	23	24	18	11

  a Find the cumulative frequencies.
  b Plot a cumulative frequency graph.
  c Find the median amount.
  d Find the upper and lower quartile amounts.

Interquartile range

The lower quartile cuts off the bottom quarter of the data. The upper quartile cuts off the top quarter of the data. In between is the middle half of the data. In the temperature example, the lower quartile was 68 and the upper quartile was 83. So the middle half of the data lay between 68 and 83.

The **interquartile range** is the difference between the quartiles. It tells us the width of the interval which contains the middle half of the data. So the interquartile range for the temperatures is 15 °F.

Interquartile range = upper quartile − lower quartile

The size of the interquartile range tells us how widely spread the data is. If it is large, then the middle half is widely spread. If it is small, then the middle half is concentrated in a small region.

Example Two groups, each of 40 students, took an exam. The diagram shows the cumulative frequency curves for their marks. Find the interquartile range for each group. What can you say about the groups?

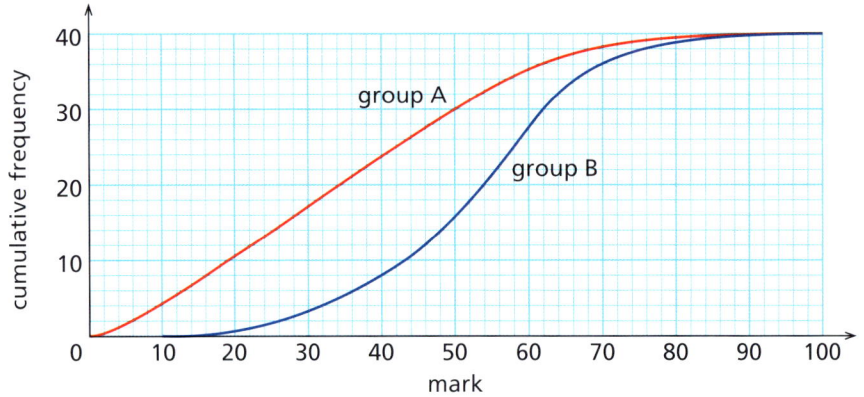

Find the quartiles by the usual process.

> For group A, lower quartile = 19 and upper quartile = 50.
> Interquartile range = 31.
> For group B, lower quartile = 43 and upper quartile = 62.
> Interquartile range = 19.

The interquartile ranges are 31 and 19.

The interquartile range is much larger for A than for B. So the middle half of group A is much more widely spread than that of group B. This means there is a wider spread of ability in group A. Notice also that the red curve is to the left of the blue curve. Hence group A's scores were lower.

The standard in group A is lower and more widely spread than in group B.

Exercise 7.6

1 The diagram shows cumulative frequency graphs for the temperatures in two places over 60 days in summer. For each place, find the interquartile range. What can you say about the two places?

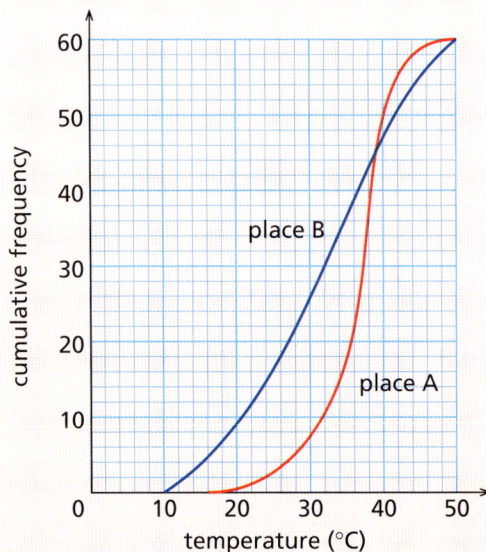

2 The diagram shows cumulative frequency graphs for the salaries of the employees in two companies. For each company, find the interquartile range. What can you say about the two companies?

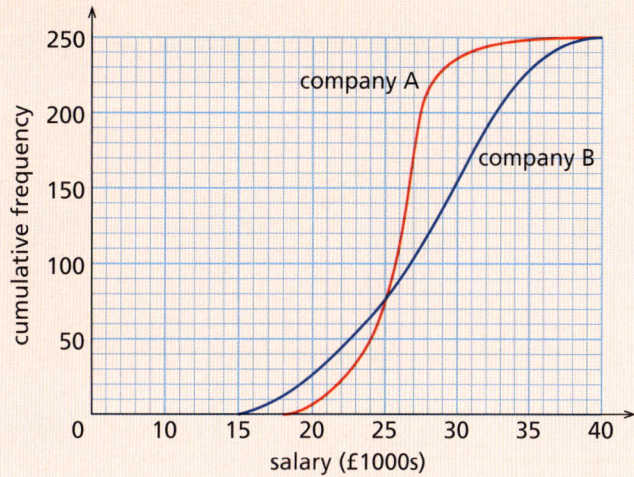

3 Two athletics clubs enter their members for a race. The diagram shows cumulative frequency graphs for the times the members of the clubs took. For each club, find the interquartile range. What can you say about the two clubs?

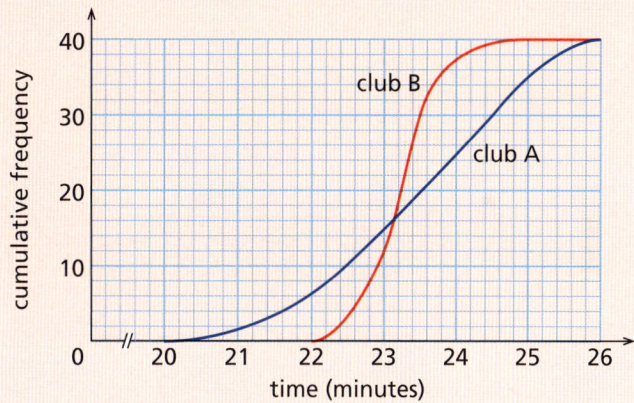

4 A factory has two machines for filling cans with cola. The diagrams shows cumulative frequency graphs for the amounts the machines dispense. For each machine, find the interquartile range. What can you say about the two machines?

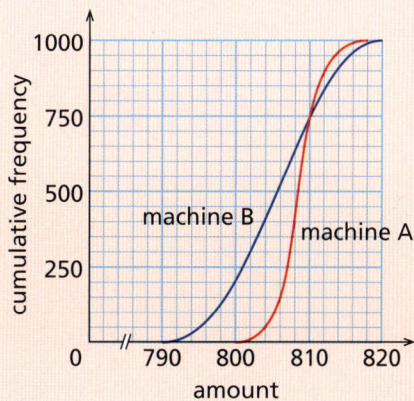

Prediction

An exam has two papers. The diagram shows the cumulative frequencies of the marks in the two papers, taken by the same group of students. Maisie got 80% in the first paper, but missed the second. What could she expect to get in the second?

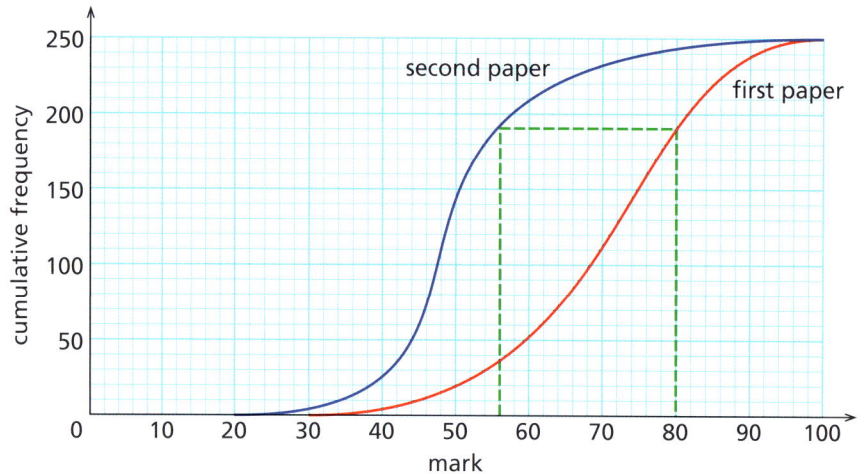

We certainly can't assume she would have got 80%. Looking at the curves, the marks in the first paper were higher, perhaps because it was easier. But it is likely that her place in the order would be about the same – if she was 60th out of 250 in the first paper, then it is likely that she would have been about 60th out of 250 in the second.

On a cumulative frequency graph, the y-coordinate gives the cumulative frequency, which corresponds to the place in the order. If points have the same y-coordinate, then they will be at the same place in the group's order.

So we can use the cumulative frequency curve to predict Maisie's mark in the second paper. Go up from 80 on the x-axis to her position in the first exam paper curve. Go across to the second exam paper curve and down to the x-axis to find her predicted mark in the second paper: 56%.

She would probably have got about 56% in the second paper.

Exercise 7.7

1 The diagram shows
   cumulative frequency curves
   for house prices in two years.
   a A house cost £90 000 in
      1999. What do you think
      it cost in 1995?
   b A house cost £60 000 in
      1995. What do you think
      it cost in 1999?

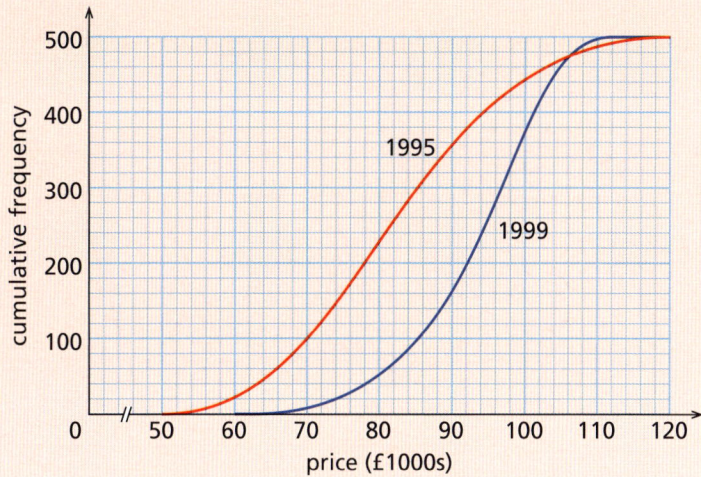

2 The diagram shows cumulative
   frequency curves for the times
   taken by the members of an
   athletics club to run a race,
   before and after an extensive
   period of training.
   a Dalbir ran the race in 53
      seconds before the
      training. What do you
      think his time was after
      the training?
   b After the training, Andy
      ran the race in 48 seconds.
      What do you think his time
      was before the training?

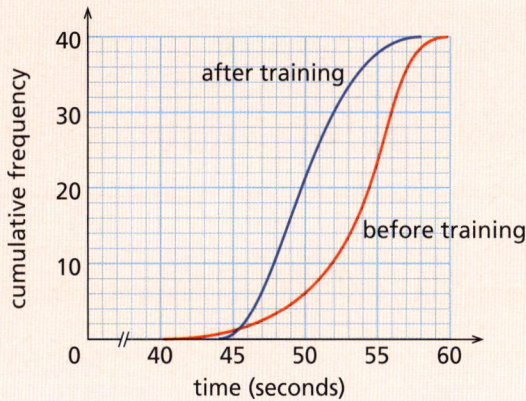

3 The diagram shows
   cumulative frequency curves
   for the marks in A level
   Maths and in GCSE Maths.
   a Anita got 64% at A level.
      What do you think she got
      at GCSE?
   b Su Ying got 66% at
      GCSE. What do you think
      she got at A level?

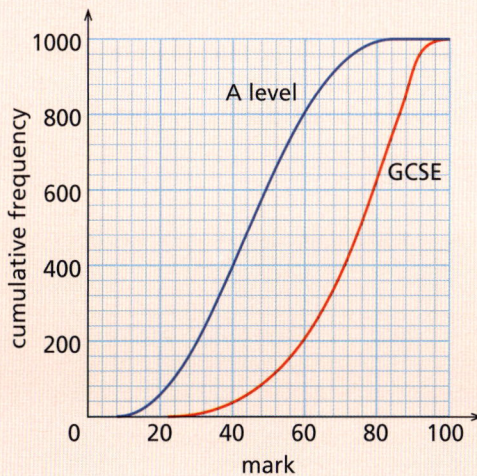

4 At the beginning of this chapter we considered Vessa, who scored 56% in the first paper of her GCSE Maths exam, but had to miss the second paper. The diagram shows cumulative frequency curves for the two papers. What should the examiners give her for the second paper?

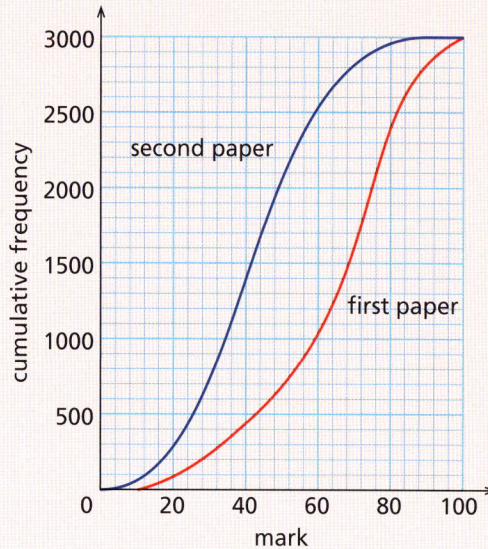

SUMMARY

■ The **cumulative frequency** at a value is the **running total** of the frequencies which are less than the value.
■ A **cumulative frequency curve** plots the cumulative frequency against the values. Be sure to plot points at the upper limits of the intervals, not at the middles.
■ The **median** can be found from the cumulative frequency graph, by finding the value whose cumulative frequency is half the total frequency.
■ The **upper and lower quartiles** are the values which cut off the top and bottom quarters of the data. They can also be found from the cumulative frequency graph.
■ The difference between the quartiles is the **interquartile range**. The size of the interquartile range is a guide to how widely spread the data is.
■ Cumulative frequency curves can be used to make predictions.

Exercise 7A

Questions 1–8 involve this frequency table, which gives the times that 100 long-distance coaches took to make a certain journey.

time (minutes)	200–	205–	210–	215–	220–225
frequency	7	36	27	22	8

1 Write out an extra row for the cumulative frequencies.
2 How many coaches took less than 215 minutes?
3 Draw a cumulative frequency graph for the data.

4 How many of the coaches took more than 207 minutes?
5 Find the median time.
6 Find the lower quartile of the times.
7 Find the upper quartile of the times.
8 Find the interquartile range of the times.

Questions 9 and 10 involve this diagram, which shows the cumulative frequency curves for the heights achieved in the high jump by the same group of athletes, before and after training.

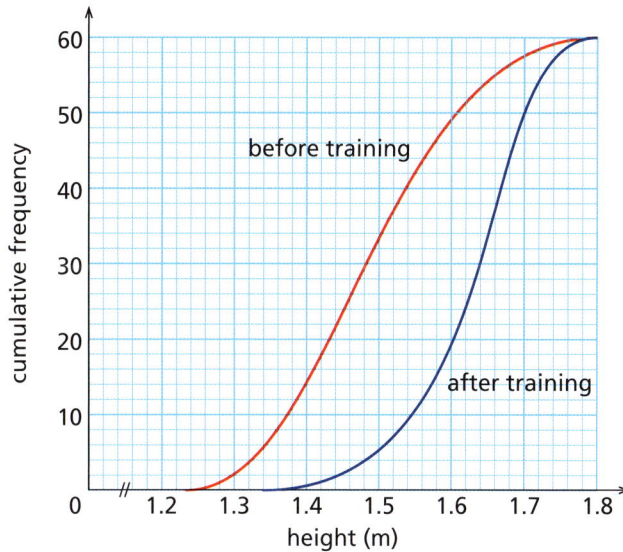

9 For each curve, find the interquartile range. What can you say about the difference between the performances?
10 Darren achieved a jump height of 1.4 m, but had to miss the training session. What height do you think he would have reached after training?

Exercise 7B

1 This table shows the amount Karim saved per week during his holiday job. Write out an extra row showing the running total.

week	1	2	3	4	5	6
amount (£)	22	17	21	12	25	32

Questions 2 to 8 involve this frequency table, which gives the times that 160 internet users spend on-line per week.

time (hours)	0–	2–	4–	6–	8–	10–	12–14
frequency	38	47	32	22	10	7	4

2 Write out an extra row for the cumulative frequencies.
3 How many users spend at least 6 hours on-line per week?
4 Draw a cumulative frequency graph for the data.

5 How many users spend less than 4.5 hours on-line?

6 Find the median of the times.

7 Find the quartile times.

8 Find the interquartile range of the times.

Questions 9 and 10 involve this diagram, which shows cumulative frequency curves for the results for a test of reaction time, taken first thing in the morning and at midday.

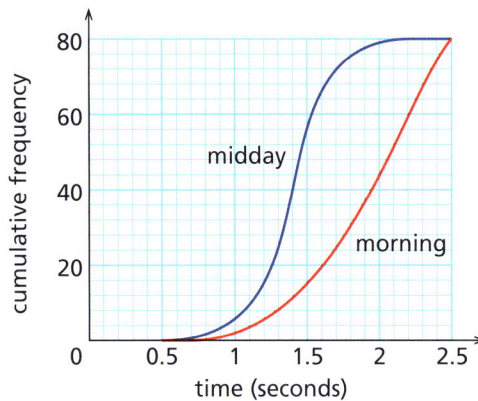

9 Find the interquartile ranges for the two curves. What is the difference between the times?

10 Susannah took 1.8 seconds in the morning test. How long do you think she took at midday?

Exercise 7C (Ma1)

In this chapter all the data has been supplied for you. Collect your own data, and draw cumulative frequency curves to illustrate it. You could investigate and compare:

> the lengths of sentences in two different magazines
> the amounts of money spent per week by two groups of people
> the amount of time per week the members of two classes spend on their homework

Exercise 7D (Ma1)

In the examples and exercises of this chapter the graphs have been carefully labelled. You can *sketch* cumulative frequency curves, however, to show the rough differences between two groups.

Suppose two groups of athletes run the same race. Sketch their cumulative frequency curves if:

> group A is, on average, faster than group B and their times are more widely spread
> group A is, on average, slower than group B and their times are more widely spread

Suppose you are given two cumulative frequency curves. Describe carefully how you can tell:

> which curve corresponds to higher numbers
> which curve corresponds to a greater spread of numbers

Exercise 7E

Plotting cumulative frequency graphs takes a lot of time. A spreadsheet can be used instead. Look at the table below, from the example on page 81.

temperature (°F)	50–	60–	70–	80–	90–100
frequency	17	46	73	52	12

Recall that we plot points at the *end* of each interval, so we want to plot at (50, 0), (60, 17) and so on. So our first frequency is 0, not 17.

In cells A1, B1 up to F1, enter 50, 60, etc. up to 100.
In cells A2, B2 up to F2, enter 0, 17, 46, etc.
In cells A3, enter =A2. In cell B3, enter =A3+B2 and copy this across to F3. The result should look like this.

	A	B	C	D	E	F	G
1	50	60	70	80	90	100	
2	0	17	46	73	52	12	
3	0	17	63	136	188	200	

Now plot the graph. Specify that you want an *xy* scatter graph, with the *x*-axis values from the first row and the *y*-axis values from the third row. Add labels if you want. The result might look like this.

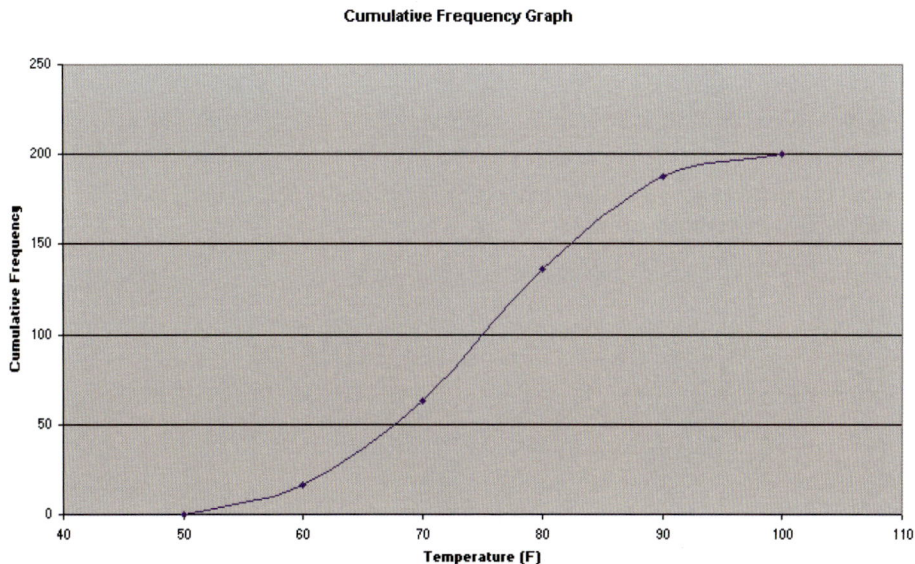

Use a computer to draw some of the cumulative frequency curves of this chapter.

8 Trigonometry

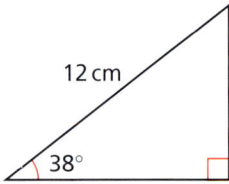

Look at the right-angled triangle on the left. We know the length of the hypotenuse, and the angles. We could draw this triangle, and then measure the other two sides. We would find that the horizontal side has length about 9.5 cm.

In the second right-angled triangle, we know all three sides. We could draw the triangle, and then measure the angles. We would find that the angle at the bottom left is about 53°.

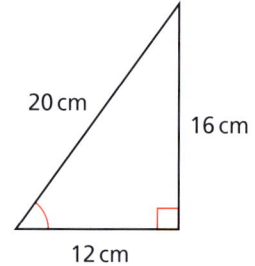

Drawing is a long and inaccurate business. If the lengths of the sides and the angles must have certain fixed values, then there must be a mathematical way to *calculate* the lengths and angles. That mathematical way involves **trigonometry**.

Labelling the sides

Trigonometry involves the sides and angles of a right-angled triangle. Before we define the functions, we must have a clear and unambiguous way of labelling the sides of the triangle. The figure below shows a right-angled triangle. The right angle is at A, and we know (or want to know) the angle at B. The three sides of the triangle are labelled as follows.

The hypotenuse (HYP) is the longest side, BC.

The opposite (OPP) is the side, AC, opposite the angle we know.

The adjacent (ADJ) is the side, AB, next to the angle we know.

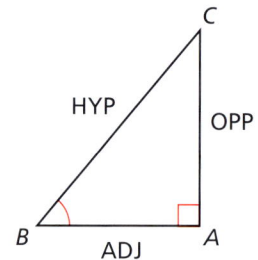

Exercise 8.1

1 Label the sides of each of these triangles with HYP, OPP and ADJ. In each case, the angle we know or want to know is marked with a red dot.

a 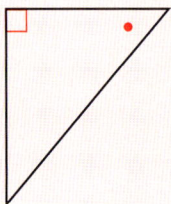   b 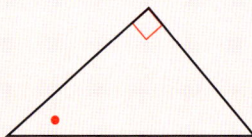   c 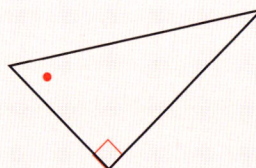   d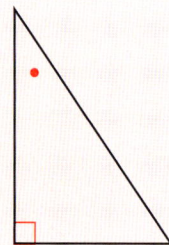

2 The diagram shows a ladder leaning against a wall. We want to know the angle between the ladder and the wall. Label the sides of the triangle with HYP, OPP and ADJ.

3 The diagram shows a ramp for wheelchair access. We want to know the angle between the ramp and the ground. Label the sides of the triangle with HYP, OPP and ADJ.

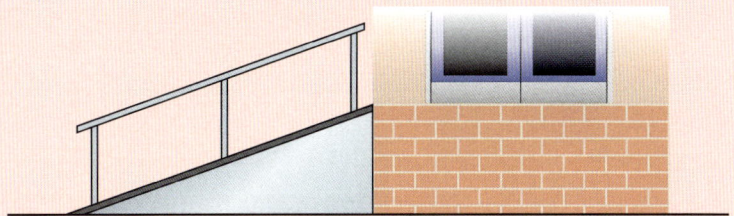

4 In triangle XYZ, $\angle XYZ = 90°$. We want to know $\angle YZX$. Draw the triangle and label the sides with HYP, OPP and ADJ.

Suppose a set of right-angled triangles are all similar to each other. Then there is a fixed ratio between the corresponding pairs of sides. In the diagram below, the triangles are all similar.

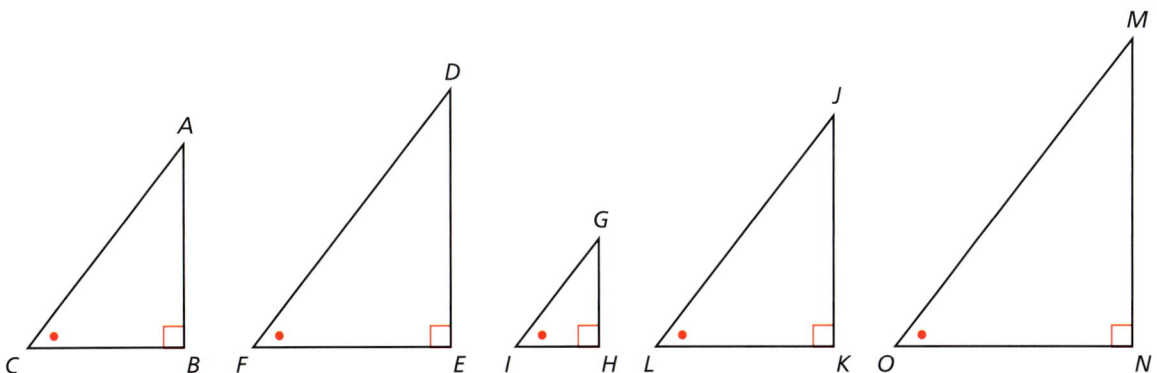

It follows that, for example, the ratio of the opposite side to the adjacent side is constant.

$$\frac{AB}{BC} = \frac{DE}{EF} = \frac{GH}{HI} = \frac{JK}{KL} = \frac{MN}{NO}$$

These ratios are all equal to the same value, called the **tangent** of the angle x. There are two other ratios, **sine** and **cosine**. The three ratios are abbreviated to **tan**, **sin** and **cos**. They are defined as follows.

$$\sin x = \frac{\text{OPP}}{\text{HYP}}$$

$$\cos x = \frac{\text{ADJ}}{\text{HYP}}$$

$$\tan x = \frac{\text{OPP}}{\text{ADJ}}$$

The ratios can be remembered by the 'words' SOHCAHTOA or OHSAHCOAT.
Sin is Opp over Hyp, Cos is Adj over Hyp, Tan is Opp over Adj
Opp over Hyp is Sin, Adj over Hyp is Cos, Opp over Adj is Tan

Example   In ABC, $\angle ABC = 90°$, $AB = 3\,\text{cm}$, $BC = 4\,\text{cm}$ and $AC = 5\,\text{cm}$. Find the sine, cosine and tangent of $\angle BAC$.

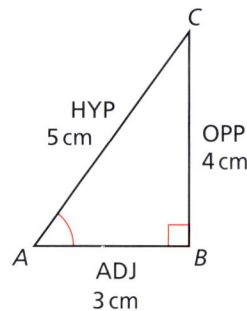

Label the sides of the triangle as shown: $AC = \text{HYP}$, $BC = \text{OPP}$ and $AB = \text{ADJ}$. Using the definitions

$$\sin x = \frac{\text{OPP}}{\text{HYP}} = \frac{4}{5}$$

$$\cos x = \frac{\text{ADJ}}{\text{HYP}} = \frac{3}{5}$$

$$\tan x = \frac{\text{OPP}}{\text{ADJ}} = \frac{4}{3}$$

The sine, cosine and tangent of $\angle BAC$ are $\frac{4}{5}$, $\frac{3}{5}$ and $\frac{4}{3}$ respectively.

Exercise 8.2

1 Write down the sine, cosine and tangent of $\angle ACB$ of the example on page 100.
2 In the triangle shown, $LM = 5\,\text{cm}$, $MN = 12\,\text{cm}$ and $LN = 13\,\text{cm}$.

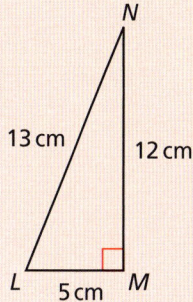

   a Use Pythagoras' theorem to check that the triangle is right-angled at M.
   b Find the sine, cosine and tangent of $\angle NLM$.
3 In the triangle shown, $PQ = 8\,\text{cm}$, $QR = 15\,\text{cm}$ and $RP = 17\,\text{cm}$.

> **Remember:**
>
> *Pythagoras' theorem*
> $c^2 = a^2 + b^2$

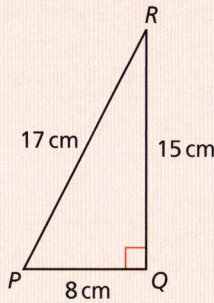

   a Check that the triangle is right-angled.
   b Find the sine, cosine and tangent of $\angle RPQ$.

The values of the three ratios can be found from a scientific calculator. For some calculators you enter the angle first, for others you enter the function first. To find $\sin 27°$, for example, one of these sequences will work.

| 2 | 7 | sin |

| sin | 2 | 7 | = |

A number beginning with $0.453\,99$ should appear.

> **Warning:** Angles can be measured in radians or in grades, as well as in degrees. Make sure that your calculator is set in degree mode. There should be a 'D' or 'deg' on the display. If you are in the wrong mode, you will get a different answer for $\sin 27$.
> In 'rad' mode, you will get $0.956\ldots$
> In 'grad' mode, you will get $0.411\ldots$

Exercise 8.3

Use your calculator to find the following. Give your answers correct to three significant figures.

1 $\sin 46°$

2 $\cos 83°$

3 $\tan 28°$

4 $\sin 12°$

5 $\cos 24°$

6 $\tan 77°$

7 $\sin 21.11°$

8 $\cos 32.77°$

9 $\tan 87.94°$

10 $\sin 22\frac{1}{2}°$

11 $\cos 77\frac{3}{4}°$

12 $\sin 15\frac{2}{3}°$

13 $\tan \frac{7}{8}°$

14 The triangle ABC has sides $AB = 3$ cm, $BC = 4$ cm and $CA = 5$ cm.

    a Construct the triangle.

    b Use a protractor to measure $\angle BAC$.

    c Use your calculator to find the sine, cosine and tangent of $\angle BAC$. Are the results close to $\frac{4}{5}$, $\frac{3}{5}$ and $\frac{4}{3}$?

15 Construct the triangle LMN of question 2 of exercise 8.2. Measure $\angle NLM$. Use your calculator to find the sine, cosine and tangent of $\angle NLM$. Are your results close to the values you obtained before?

Finding sides

If you have a right-angled triangle in which you know one other angle and one of the sides, then you can use trigonometry to find the other sides. Look at the definitions of sine, cosine and tangent.

$$\sin x = \frac{\text{OPP}}{\text{HYP}}$$

$$\cos x = \frac{\text{ADJ}}{\text{HYP}}$$

$$\tan x = \frac{\text{OPP}}{\text{ADJ}}$$

Multiplying across,

$$\text{HYP} \times \sin x = \text{OPP}$$
$$\text{HYP} \times \cos x = \text{ADJ}$$
$$\text{ADJ} \times \tan x = \text{OPP}$$

So, for example, if you know the angle x and the ADJ side you can use the tan function to find the OPP side.

Example   In the triangle shown, $\angle ABC = 90°$, $\angle BAC = 37°$ and $AC = 20$ cm. Use the sine function to find BC.

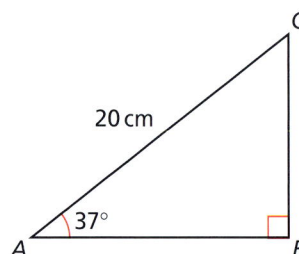

We are given AC, which is the HYP of the triangle. The side BC is the OPP side. Using the formula

$$OPP = HYP \times \sin x$$
$$BC = 20 \times \sin 37°$$
$$= 20 \times 0.6018$$
$$= 12.036$$

$BC = 12.04$ cm.

Exercise 8.4

1 Find the unknown sides of these triangles. In each case use the function given.

a sin 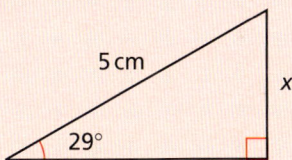 **b** tan 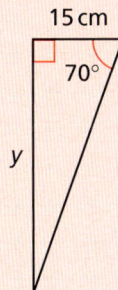 **c** cos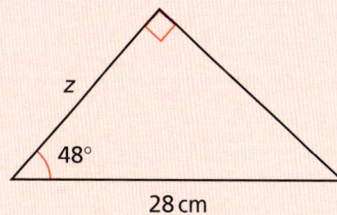

d tan 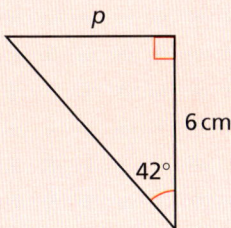 **e** cos 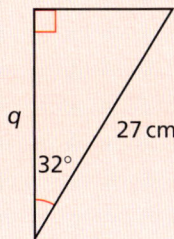 **f** sin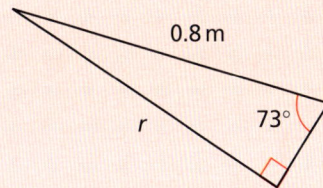

2 In $\triangle LMN$, $\angle MLN = 90°$, $\angle LMN = 67°$ and $ML = 5$ cm. Use the tangent function to find LN.
3 In $\triangle PQR$, $\angle PQR = 90°$, $\angle PRQ = 49°$ and $PR = 6$ m. Use the sine function to find PQ.
4 In $\triangle XYZ$, $\angle XZY = 90°$, $\angle XYZ = 37°$ and $XY = 40$ mm. Use the cosine function to find YZ.
5 In $\triangle ABC$, $\angle BAC = 90°$.
  a If $BC = 8$ m and $\angle CBA = 67°$ find BA.   (Use cosine.)
  b If $BC = 20$ cm and $\angle BCA = 49°$ find AB.   (Use sine.)
  c If $AB = 12$ mm and $\angle CBA = 77°$ find AC.   (Use tangent.)

In the previous exercise you were told which function to use. Here you must decide which function is needed. Label the sides of the triangle with HYP, OPP and ADJ. There is one side you know, and one side you want to know. Look at the definitions to decide which function should be used.

If the sides are OPP and HYP then use sine.
If the sides are ADJ and HYP then use cosine.
If the sides are OPP and ADJ then use tangent.

Example In triangle PQR, $\angle P = 90°$, $\angle R = 48°$ and $QR = 40\,\text{cm}$. Find PR.

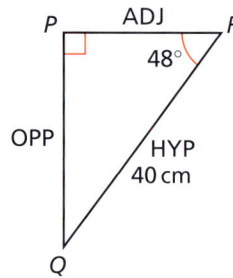

Label the sides of the triangle. QR is HYP, PR is ADJ and QP is OPP. The side we know and the side we want to know are HYP and ADJ. The function which links these sides is cosine.

$$\cos x = \frac{\text{ADJ}}{\text{HYP}}$$

Hence $40 \times \cos 48° = PR$.

$$PR = 40 \times 0.669\,13$$
$$= 26.765$$

$PR = 26.8\,\text{cm}$.

Exercise 8.5

Remember:
SOHCAHTOA

1 Find the unknown sides of these triangles.

a

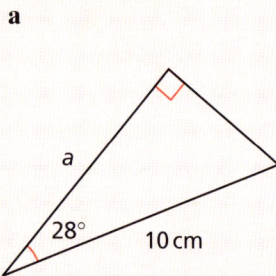

b

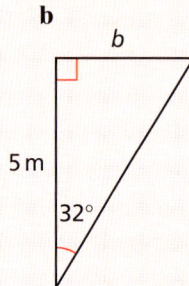

c

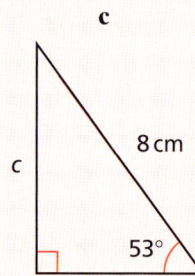

d

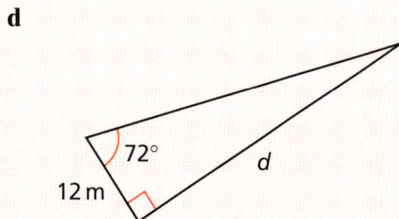

e

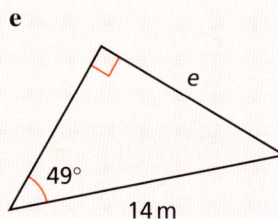

2 For each of the following, sketch the triangle using the details given. Find the unknown side.
   a $\angle ABC = 90°$, $AC = 12\,cm$, $\angle ACB = 74°$. Find BC.
   b $\angle PQR = 90°$, $PQ = 4\,m$, $\angle RPQ = 28°$. Find RQ.
   c $\angle LMN = 90°$, $LN = 66\,mm$, $\angle LNM = 41°$. Find LM.
   d $\angle XYZ = 90°$, $XZ = 42\,km$, $\angle ZXY = 32°$. Find YX.
   e $\angle DEF = 90°$, $DE = 42\,m$, $\angle FDE = 64°$. Find EF.
   f $\angle JKL = 90°$, $JL = 5$ miles, $\angle LJK = 41°$. Find LK.

Finding angles

If you know one other angle of a right-angled triangle and one side, then you can find the other sides. If you know two sides of a right-angled triangle, then you can find the angles.

The functions sine, cosine and tangent go from the angle to the ratio of the sides. There are corresponding functions which go from the ratio of the sides to the angle. They are called $\sin^{-1}$, $\cos^{-1}$ and $\tan^{-1}$, or **arcsin**, **arcos** and **arctan**.

Note. $\sin^{-1} r$ is not the same as $\dfrac{1}{\sin r}$

$$x = \sin^{-1} \frac{OPP}{HYP}$$

$$x = \cos^{-1} \frac{ADJ}{HYP}$$

$$x = \tan^{-1} \frac{OPP}{ADJ}$$

A calculator can evaluate these functions. Usually you must press 'inv', 'shift' or 'second function' first. The sequence to find $\sin^{-1} 0.47$ might be

| shift | sin | . | 4 | 7 | = |

| . | 4 | 7 | shift | sin |

An answer beginning 28.034 should appear.

Note. Make sure your calculator is set in degree mode. Also, be careful with fractions, especially if, for your calculator, you put the function first. To find $\sin^{-1}\frac{1}{2}$, for example, either use brackets or write $\frac{1}{2}$ as 0.5.

| shift | sin | (| 1 | ÷ | 2 |) | = |

The answer 30 should appear. If you don't use brackets, as in

| shift | sin | 1 | ÷ | 2 | = |

you will get the incorrect answer 45°.

Exercise 8.6

Find the following.

1 $\sin^{-1} 0.3$ 2 $\cos^{-1} 0.6$
3 $\tan^{-1} 1.8$ 4 $\sin^{-1} 1\frac{1}{3}$
5 $\cos^{-1} 1\frac{1}{4}$ 6 $\tan^{-1} 1\frac{2}{3}$
7 $\sin^{-1} \frac{2}{7}$ 8 $\cos^{-1} \frac{4}{9}$
9 $\tan^{-1} 1\frac{1}{4}$ 10 $\tan^{-1} 2\frac{3}{8}$

11 In triangle ABC, $\angle ABC = 90°$, $AC = 5$ cm, $AB = 3$ cm and $BC = 4$ cm. Find $\angle BAC$ as:
   a $\sin^{-1} \frac{4}{5}$ b $\cos^{-1} \frac{3}{5}$ c $\tan^{-1} \frac{4}{3}$
   Check that your answers are the same.

12 In triangle LMN, $\angle LMN = 90°$, $LN = 17$ cm, $LM = 15$ cm and $MN = 8$ cm. Find $\angle MNL$ as:
   a $\sin^{-1} \frac{15}{17}$ b $\cos^{-1} \frac{8}{17}$ c $\tan^{-1} 1\frac{7}{8}$

13 On page 105, we stated that if you press this sequence for $\sin^{-1} \frac{1}{2}$

   [shift] [sin] [1] [÷] [2] [=]

   you will get the incorrect answer 45°. Explain why.

Suppose you are given two sides of a right-angled triangle. Then you can find the angles of the triangle. Make sure you use the right ratio. Label the sides of the triangle and decide which ratio to use.

Example   In triangle PQR, $\angle RPQ = 90°$, $PQ = 7$ cm and $PR = 12$ cm. Find $\angle PRQ$.

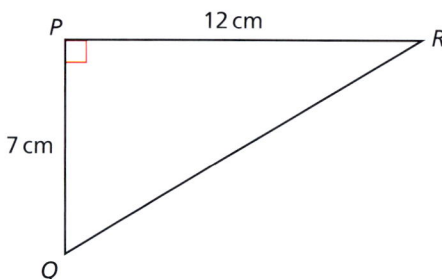

PQ is the OPP, and PR is the ADJ. The function which links OPP and ADJ is tangent.

$$\tan R = \frac{\text{OPP}}{\text{ADJ}} = \frac{7}{12}$$

So $\angle PRQ = \tan^{-1} \frac{7}{12} = \tan^{-1} 0.5833 = 30.3°$.

$$\angle PRQ = 30.3°$$

Exercise 8.7

1 Find the unknown angles in these triangles.

a

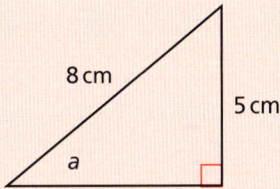

8 cm
5 cm
a

b

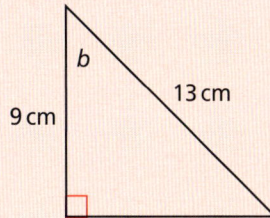

b
13 cm
9 cm

c

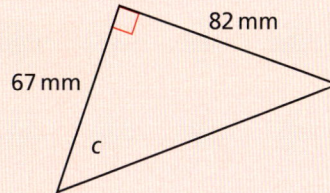

82 mm
67 mm
c

d

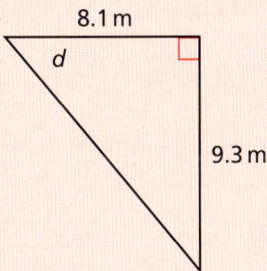

8.1 m
d
9.3 m

e

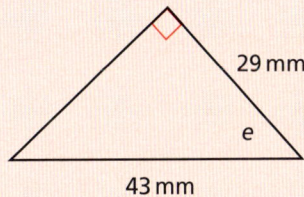

29 mm
e
43 mm

f

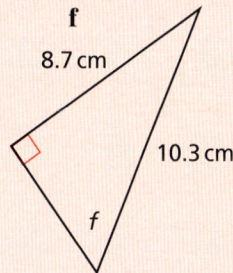

8.7 cm
10.3 cm
f

2 In the following questions, sketch the triangle described and find the unknown angle.
  a In $\triangle ABC$, $\angle ABC = 90°$, $AB = 4.3$ cm and $BC = 6.7$ cm. Find $\angle BAC$.
  b In $\triangle LMN$, $\angle MNL = 90°$, $ML = 5$ m and $LN = 3.5$ m. Find $\angle LMN$.
  c In $\triangle PQR$, $\angle PRQ = 90°$, $PR = 47$ mm and $PQ = 66$ mm. Find $\angle RPQ$.
  d In $\triangle DEF$, $\angle DEF = 90°$, $DF = 32$ m and $EF = 17$ m. Find $\angle DFE$.
  e In $\triangle XYZ$, $\angle XYZ = 90°$, $XY = 26$ mm and $YZ = 19$ mm. Find $\angle YZX$.
  f In $\triangle IJK$, $\angle IJK = 90°$, $IK = 0.5$ km and $IJ = 0.35$ km. Find $\angle JIK$.

Practical uses

Trigonometry has many practical uses. First, there are some important words to describe angles.

When we look up at something, the angle between our line of sight and the horizontal is the **angle of elevation**.

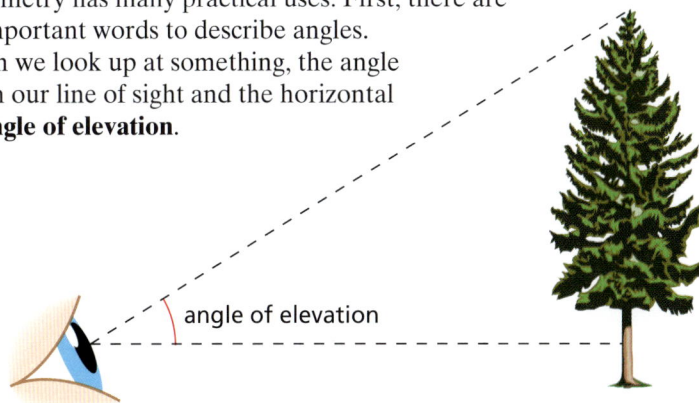

angle of elevation

When we look down at something, the angle between our line of sight and the horizontal is the **angle of depression**.

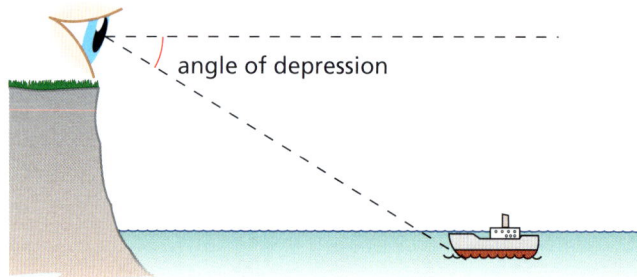

angle of depression

Notice that both angles are measured from the horizontal, not the vertical.

When a ship or plane is travelling, its direction is given by a **bearing**. The bearing is the angle its direction makes with North, measured clockwise.

Remember:

We always write a bearing with three digits.

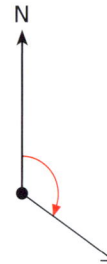

N

Examples   A beam of length 3.5 m leans against a wall, reaching 3 m up it. What is the angle between the beam and the ground?

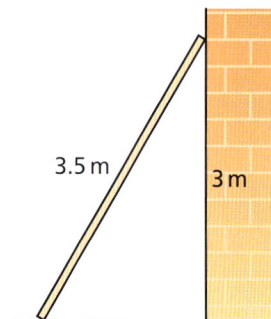

3.5 m

3 m

The length of the beam is the HYP of the triangle. The length up the wall is the OPP. The function to use is sine. If the angle with the horizontal is x, then

$$\sin x = \frac{\text{OPP}}{\text{HYP}} = \frac{3}{3.5}$$

Hence $x = \sin^{-1}\left(\dfrac{3}{3.5}\right) = 59°$.

The beam makes $59°$ with the horizontal.

A ship travels 150 km along a bearing of 235°. How far South has it travelled?

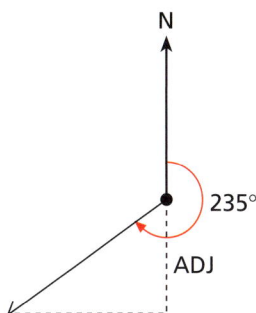

The bearing is measured from North. The direction of the ship makes 55° with South. The journey of the ship is the HYP, and the South part of the journey is the ADJ. The function to use is cos.

$$\cos 55° = \frac{ADJ}{HYP} = \frac{ADJ}{150}$$

Hence ADJ = $150 \times \cos 55°$
$\qquad\qquad = 150 \times 0.5736$
$\qquad\qquad = 86.04$

The ship has travelled 86 km South.

Exercise 8.8

1 At a distance of 60 m from the base of a tower, the angle of elevation of its top is 29°. How high is the tower?

2 A 3 m girder leans against a wall, at an angle of 68° to the horizontal.
   a How high up the wall does the girder reach?
   b How far is the foot of the girder from the base of the wall?

3 A 2.5 m ladder leans against a wall, reaching 2.1 m up it. What is the angle between the ladder and the wall?

4 A point is 50 m from the base of a 60 m tower. From the point, what is the angle of elevation of the top of the tower?

5 A tower is 80 m high. From the top of the tower, what is the angle of depression of a point which is 70 m from the base of the tower?

6 A boat is 300 m out to sea. The angle of depression of the boat from the top of a cliff is 12°. How high is the cliff?

7 A rectangle has sides 6 m and 11 m. What is the angle between the diagonal and the longer side?

8 A ship sails for 150 km on a bearing of 067°.
   a How far North has it travelled?   **b** How far East has it travelled?

9 A section of road is 250 m long. If it rises for 15 m, find the angle between the road and the horizontal.

10 A clock face is vertical. The hour hand is 0.4 m long. How high is the tip of the hand above the centre of the clock at the following times?
   a 3 o'clock   **b** 1 o'clock   **c** 4 o'clock

11 The diagonals of a rhombus are 6 cm and 14 cm. Find the angles of the rhombus.

12 A road slopes at 7° to the horizontal. After walking 300 m along the road, how far have I travelled:
   a horizontally?   **b** vertically?

13 An aircraft is spotted directly overhead, flying horizontally. Ten seconds later it has travelled 900 m and its angle of elevation is 72°. How high is the plane flying?

14 Marie stands by a straight canal and sees a tree directly opposite on the other bank. She walks 20 m along the canal, and the line from her to the tree now makes 47° with the bank. How wide is the canal?

15 At the *Onbashira* festival in central Japan, huge tree trunks are dragged from the forest to a shrine. At one stage the trunks are slid down a steep slope. Young men ride on the trunks, trying to stay on for as long as possible. If the angle of the slope is 35°, and the length of the slope is 80 m, find the vertical distance through which the trunks fall.

Dividing by the ratio

In the definitions of sine and cosine, the HYP side was the denominator of the fraction. So to find the other sides, OPP or ADJ, you multiply by the HYP. In all the examples of sine or cosine so far, you were given the hypotenuse. If instead you are given the OPP or the ADJ, and you want to find the hypotenuse, you need to divide by the sine or the cosine instead of multiplying by it.

Similarly, if you are given the OPP and want to find the ADJ, you divide by the tangent.

$$\sin x = \frac{OPP}{HYP} \quad \text{hence} \quad HYP \times \sin x = OPP$$

This gives $HYP = \dfrac{OPP}{\sin x}$.

Similarly,

$$HYP = \frac{ADJ}{\cos x} \quad \text{and} \quad ADJ = \frac{OPP}{\tan x}$$

Examples In $\triangle ABC$, $\angle BAC = 90°$, $\angle B = 63°$ and $AC = 12$ cm. Find BC.

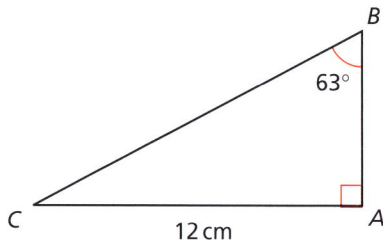

Here BC is the HYP and AC is the OPP. The function to use is sine.

$$\sin 63° = \frac{12}{HYP} \quad \text{hence} \quad HYP = \frac{12}{\sin 63°}$$

$$HYP = 12 \div 0.891\,00 = 13.468$$

$BC = 13.5$ cm.

In $\triangle XYZ$, $\angle XYZ = 90°$, $\angle YXZ = 26°$ and $YZ = 1.5\,\text{m}$. Find XY.

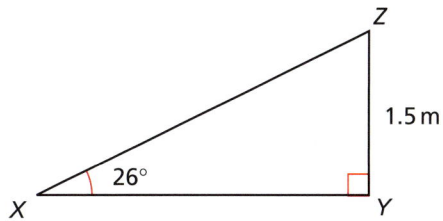

Here YZ is the OPP and XY is the ADJ. The function to use is tangent.

$$\tan 26° = \frac{\text{OPP}}{\text{ADJ}} = \frac{1.5}{XY}$$

Multiply by XY and divide by $\tan 26°$.

$$XY = \frac{1.5}{\tan 26°} = 3.075$$

$XY = 3.08\,\text{m}$.

Note. By subtraction, $\angle XZY = 64°$. We could multiply by $\tan 64°$ instead of dividing by $\tan 26°$.

$$XY = 1.5 \times \tan 64° = 3.075$$

Exercise 8.9

1 Find the unknown sides in the triangles shown.

a

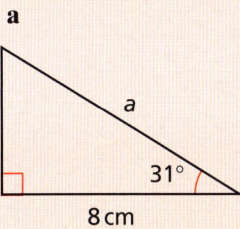

b

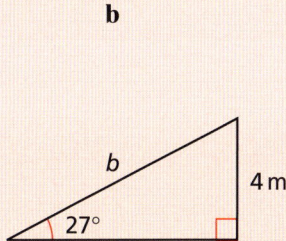

c

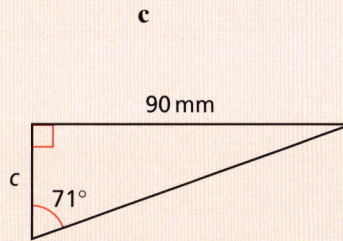

d

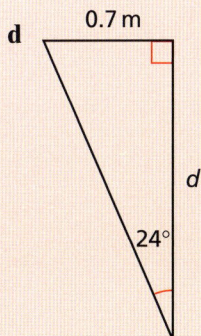

e

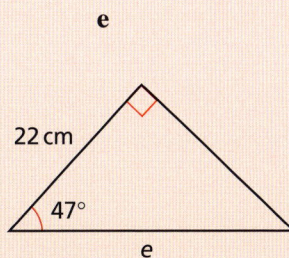

f

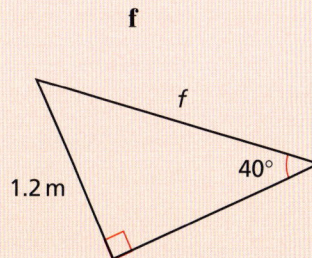

2 In the following, sketch the triangles and find the unknown sides.
   a In $\triangle ABC$, $\angle BAC = 90°$, $AB = 4$ cm and $\angle ABC = 63°$. Find BC.
   b In $\triangle PQR$, $\angle PRQ = 90°$, $\angle PQR = 53°$ and $PR = 7$ m. Find PQ.
   c In $\triangle LMN$, $\angle MNL = 90°$, $\angle MLN = 85°$ and $MN = 0.8$ m. Find LN.
   d In $\triangle XYZ$, $\angle XYZ = 90°$, $XY = 0.7$ m and $\angle YXZ = 47°$. Find XZ.
   e In $\triangle DEF$, $\angle FDE = 90°$, $\angle DEF = 66°$ and $FD = 64$ mm. Find FE.
   f In $\triangle HIJ$, $\angle IJH = 90°$, $\angle JHI = 71°$ and $JI = 0.32$ km. Find JH.

3 After travelling on a bearing of $035°$, a ship is 20 km North of its starting point. How far has it travelled?

4 A plane flies on a bearing of $124°$, ending up 40 km East of its starting point. How far has it flown?

5 The shorter side of a rectangle is 5 cm, and the diagonal makes $40°$ with the longer side. Find the length of the diagonal.

6 A horizontal plank is 5 cm thick. A thin hole is drilled through it at $20°$ to the vertical. What is the length of the hole?

7 From the top of a 150 m cliff two boats can be seen due West. Their angles of depression are $13°$ and $8°$. Find the distance between the boats.

8 A tower is 39 m high. Point A is due North of the tower, and from A the angle of elevation of the top is $28°$. Point B is due South of the tower, and from B the angle of elevation of the top is $32°$. Find the distance AB.

9 A kite is 34 m above the ground. The string of the kite, which is in a straight line, makes $46°$ with the horizontal. What is the length of the string?

10 A ladder leans against a wall, making $77°$ with the horizontal. If the ladder reaches 2.5 m up the wall, find the length of the ladder.

11 A path slopes at $11°$ to the horizontal. How far do you have to walk along the path to rise a distance of 3 m?

12 A stick is stuck in the ground at an angle of $22°$ to the horizontal. The top of the stick is 2.3 m above the ground. How long is the stick?

Miscellaneous examples

Examples $\triangle ABC$ is isosceles, with $BC = 8$ cm and $\angle ABC = \angle ACB = 80°$. Find AC.

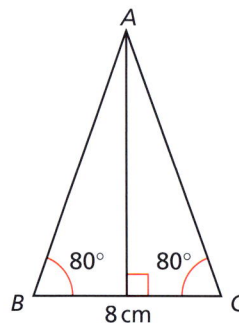

Drop a perpendicular from A to D on BC. Then $DC = 4$ cm. Use the cosine function in $\triangle DAC$.

$$\cos 80° = \frac{4}{AC}$$

$$AC = \frac{4}{\cos 80°} = 23.0$$

$AC = 23.0$ cm.

In the diagram, $\angle ACB = \angle ADC = 90°$, $\angle ABC = 64°$, $\angle DAC = 25°$ and $AB = 12\,\text{cm}$. Find AD.

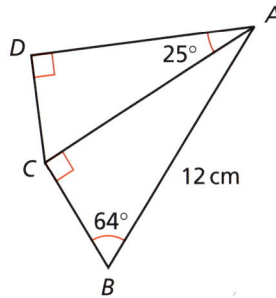

First find AC. Use the sine function in $\triangle ABC$.

$$AC = 12 \times \sin 64°$$
$$= 10.8\,\text{cm}$$

Now use the cosine function in $\triangle ACD$.

$$AD = AC \times \cos 25°$$
$$= 9.78$$

$AD = 9.78\,\text{cm}$.

Note. In this example we need the result of the first calculation to use in the second. Put the result of the first calculation in the calculator's memory. Write down the rounded value of AC, 10.8, but for the second calculation use the more accurate value in the memory. If you use the rounded value of AC, then the final value of AD is not correct to three significant figures.

Exercise 8.10

1 Find the unknown sides in these triangles.

a

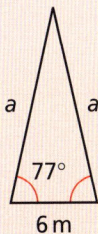

b

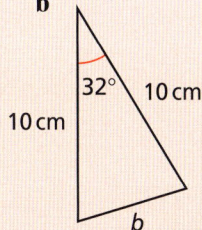

c

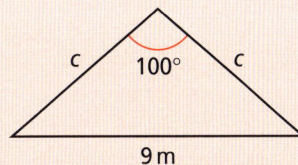

2 Find the unknown angles in these triangles.

a

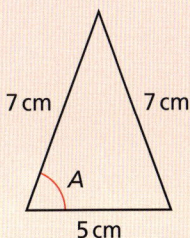

b

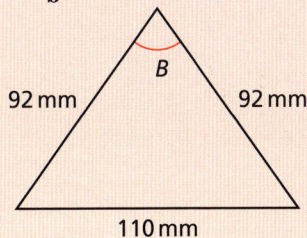

c

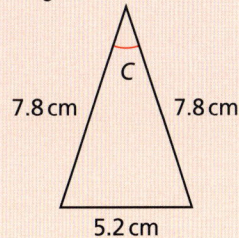

3 In each of the following, sketch the triangle and find the required side or angle.
  a In $\triangle ABC$, $AB = BC = 44$ cm and $\angle ABC = 66°$. Find AC.
  b In $\triangle PQR$, $PQ = QR = 5$ cm and $PR = 4$ cm. Find $\angle PRQ$.
  c In $\triangle LMN$, $\angle MNL = \angle LMN = 52°$ and $MN = 6.2$ cm. Find LM.
  d In $\triangle DEF$, $DE = EF = 56$ mm and $DF = 47$ mm. Find $\angle DEF$.
4 In $\triangle ABC$, $BC = 10$ cm and $\angle ABC = \angle ACB = 41°$. Find the area of the triangle.
5 The sides of a step ladder are 2.2 m long. When the sides are separated by 32° find
  a the distance apart of the bases of the sides
  b the height of the top of the ladder above the ground.

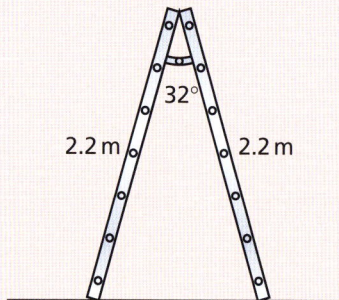

6 Two equal length planks lean against each other. Each plank makes 58° with the ground, and their feet are 2 m apart. Find
  a the length of each plank
  b the highest point of each plank above the ground.

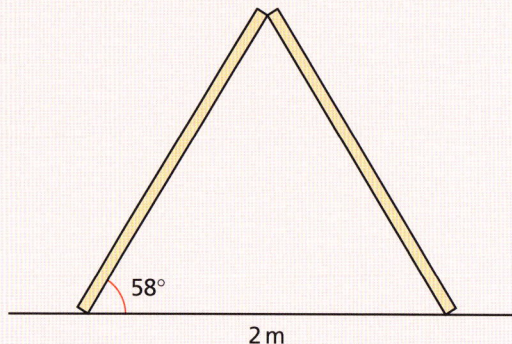

7 The maximum separation of a pair of compasses is 46°. If each side has length 8 cm, find the greatest radius of a circle that can be drawn with the compasses.
8 Find the length of RS in the diagram shown.

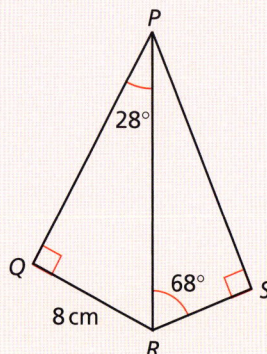

9 Find $\angle XZY$ in the diagram shown.

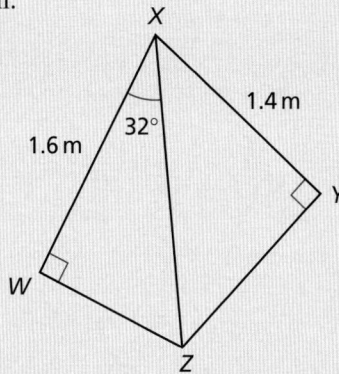

10 Find BD in the diagram shown.

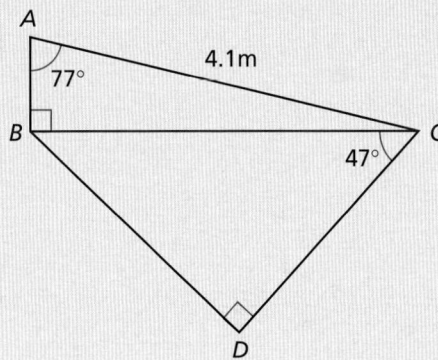

11 Find $\angle SQR$ in the diagram shown.

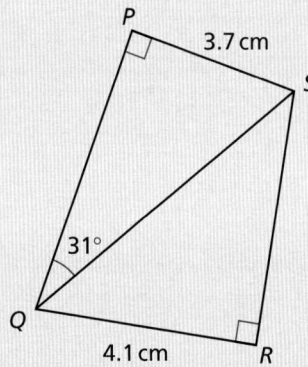

12 Find $\angle ABC$ in the diagram shown.

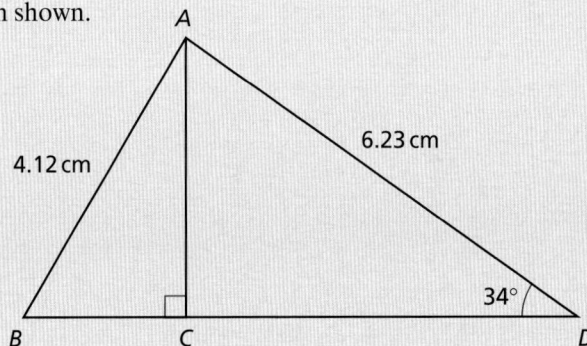

SUMMARY

■ **Trigonometry** involves the sides and angles of a right-angled triangle.
■ The longest side of a right-angled triangle is the hypotenuse, or HYP. The side opposite the angle we are dealing with is the opposite, or OPP, and the side next to the angle we are dealing with is the adjacent, or ADJ.
■ The three functions **sine**, **cosine** and **tangent** are defined by

$$\sin x = \frac{\text{OPP}}{\text{HYP}} \qquad \cos x = \frac{\text{ADJ}}{\text{HYP}} \qquad \tan x = \frac{\text{OPP}}{\text{ADJ}}$$

■ If you know one other angle of a right-angled triangle and one side, then you can find the other sides. For example, if you know the hypotenuse is 10 m and the angle is 40°, then the OPP side is $10 \text{ m} \times \sin 40°$, which is 6.43 m.
■ If you know two sides of a right-angled triangle, then you can find the angles. For example, if you know the OPP is 3 m and the ADJ is 8 m, then the angle is $\tan^{-1} \frac{3}{8}$, which is 20.6°.
■ Trigonometry has many practical uses.
■ When we look up at something, the angle between our line of sight and the horizontal is the **angle of elevation**.
■ When we look down at something, the angle between our line of sight and the horizontal is the **angle of depression**.

Exercise 8A

1 Find the unknown side x in this triangle.

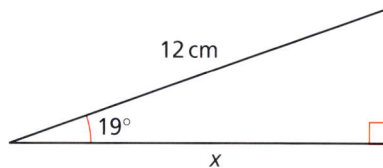

2 Sketch $\triangle ABC$, in which $\angle BCA = 90°$, $AB = 9$ cm and $\angle ABC = 55°$. Find AC.
3 Find $\tan^{-1} 2\frac{1}{7}$.
4 Find the unknown angle in this triangle.

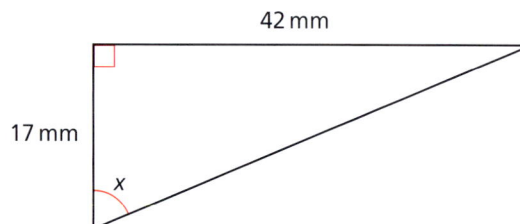

5 Sketch $\triangle LMN$, in which $\angle MLN = 90°$, $LN = 6$ m and $LM = 7$ m. Find $\angle LNM$.

6 In this diagram, $AB = 8\,\text{cm}$, $\angle ABC = 73°$, $\angle DAC = 34°$ and AD is perpendicular to BC. Find DC.

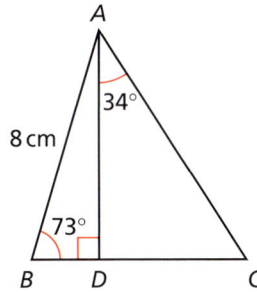

7 A ship sails to a point that is 40 km North and 75 km East of its starting point. On what bearing has it been sailing?

8 A pole is held vertical by a straight wire 37 m long at 27° to the horizontal. How high is the pole?

9 Find the unknown side x of this triangle.

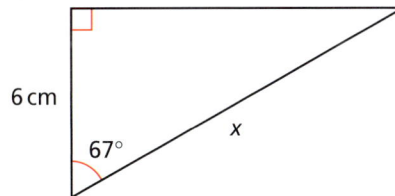

10 A plane is flying 4000 feet high. The angle of elevation of the plane from an observer on the ground is 28°. What is the distance between the plane and the observer?

Exercise 8B

1 Find the unknown side x in this triangle.

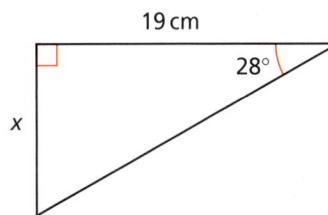

2 Sketch $\triangle XYZ$, in which $\angle YZX = 90°$, $YX = 73\,\text{m}$ and $\angle XYZ = 36°$. Find YZ.

3 Find $\cos^{-1}\frac{1}{12}$.

4 Find the unknown angle in this triangle.

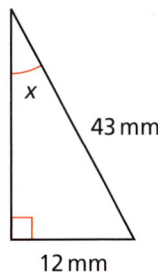

5 Sketch $\triangle HIJ$, in which $\angle IHJ = 90°$, $IJ = 50\,\text{mm}$ and $HJ = 17\,\text{mm}$. Find $\angle HJI$.

6 A roof slopes at 28° to the horizontal. If it is 3 m from the gutter to the ridge, how high is the ridge above the gutter?

7 A tree is 21 m high. The distance from the top of the tree to a point on the ground is 62 m. What is the angle of depression of the point from the top of the tree?

8 A triangle has sides 8 cm, 8 cm and 5 cm. Find the angles of the triangle.

9 In the diagram, $\angle ABC = \angle ADC = 90°$ and $\angle BAC = 49°$. $AB = 2.2\,\text{m}$ and $DC = 1.8\,\text{m}$. Find $\angle DAC$.

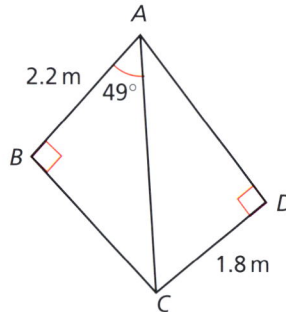

10 A mine shaft descends at an angle of 37° to the horizontal. If the bottom of the shaft is 200 m below the top, how long is the shaft?

Exercise 8C

On 11 August 1999 there was a total eclipse, visible (just!) in South West England. There will not be another total eclipse visible in the UK until 2090.

A solar eclipse occurs when the Moon comes between the Sun and the Earth. Sometimes the Moon completely covers the Sun; sometimes there is a bright ring surrounding the Moon (an **annular eclipse**). Here are some details about the Sun and Moon. Distances are in kilometres.

	diameter	least distance from Earth	greatest distance from Earth
Sun	1.392×10^6	1.466×10^8	1.526×10^8
Moon	3476	363 300	405 500

1 When will the Moon completely cover the Sun?

2 When will there be an annular eclipse?

3 Which is more likely, an annular eclipse or a total covering?

Exercise 8D

A scientific calculator can handle angles in grades, radians and degrees. At this stage you use only degrees, but if you go on to A level Maths you will meet radians.

1 Use your calculator to find how many grades there are in a right angle.
2 Use your calculator to find how many radians there are in a right angle, and hence how many radians are equivalent to 180°. Does this number look familiar?
3 Grades are part of the metric system, invented during the French Revolution, but they never became widely adopted. What are the advantages and disadvantages of measuring angles in grades instead of degrees?

Exercise 8E

A spreadsheet has functions to calculate sine, cosine and tangent. You can use these functions to draw graphs. The functions are defined for any angle, not just between 0° and 90°.

In A1 enter 0. In A2 enter =A1+5, and copy this formula down to A145. You will have angles from 0° to 720°, in steps of 5°.

The spreadsheet function will measure angles in radians rather than degrees. The formula to find the sine of the angle in A1 is =SIN(A1*PI()/180). Put this formula in B1. Copy down to B145.

	A	B	C	D	E	F	G
1	0	0					
2	5	0.087156					
3	10	0.173648					
4	15	0.258819					
5	20	0.34202					
6	25	0.422618					
7	30	0.5					

Plot the graph of the function, taking the A column as the *x*-axis and the B column as the *y*-axis.

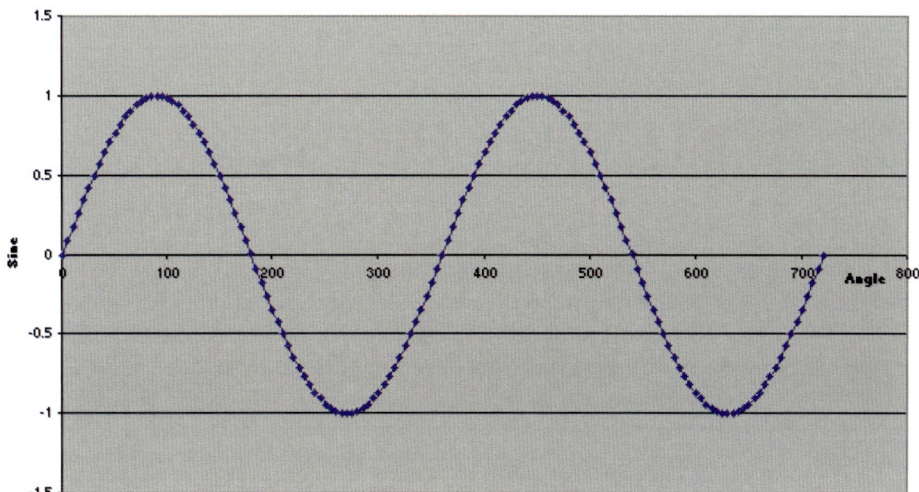

Sine Wave

Where does this sort of curve occur in the natural world?

9 Inequalities

An equation states that one expression is equal to another. By contrast, an **inequality** states that one expression may be different from another. There are four symbols to describe inequality.

$$x < y \quad x \text{ is less than } y$$
$$x > y \quad x \text{ is greater than } y$$
$$x \leq y \quad x \text{ is less than or equal to } y$$
$$x \geq y \quad x \text{ is greater than or equal to } y$$

There are many other ways of expressing these inequalities in words. For example,

$$x \leq y \quad x \text{ is at most } y, \text{ or } x \text{ is not greater than } y$$
$$x \geq y \quad x \text{ is at least } y, \text{ or } x \text{ is not less than } y$$

Exercise 9.1

1 Which of these statements are true?
   a $5 < 6$ b $6 > 5.9$ c $3\frac{1}{2} > 3.5$ d $1.25 \geq 1\frac{1}{4}$
   e $-2 < 7$ f $-8 \geq -1$ g $6 \geq -12$ h $-0.5 \geq -\frac{1}{2}$
   i $17 > 17$ j $17 \leq 17$ k $\frac{4}{3} \leq 1.3$ l $1.2 > \frac{6}{5}$

2 Write each of the following statements using one of the four inequality symbols. Hence find which statements are equivalent to each other.
   a x is larger than 3. b x is at most 3.
   c 3 is smaller than or equal to x. d 3 is at most x.
   e x is greater than or equal to 3. f 3 is less than x.
   g x is not less than 3. h 3 is not greater than x.

3 Below are some statements about money. If I have £x, write each statement as an inequality in x.
   a I have at least £10. b I have less than £20.
   c I don't have more than £15. d I have at most £12.
   e I don't have less than £14. f I have more than £5.
   g I can afford a £7 book. h I can't afford a £25 ticket.

4 What is wrong with the following argument?

   Nothing is better than Heaven.
   Half a loaf is better than nothing.

Hence

   Half a loaf is better than Heaven.

> $0 >$ Heaven
> Half a loaf > 0
>
> Half a loaf $>$ Heaven

Illustrating inequalities on the number line

Inequalities can be shown on a **number line**. The numbers obeying the inequalities are shown by a bar above the line. If the inequality involves $<$ or $>$, the bar ends with a hollow dot. If it involves $\leq$ or $\geq$, then it ends with a solid dot. Below are shown the lines corresponding to $x < 3$ and $x \geq -2$.

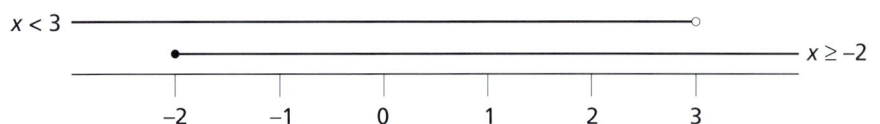

Exercise 9.2

1 Write down the inequalities illustrated below.

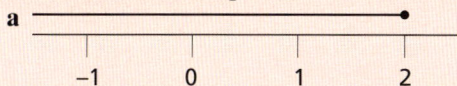

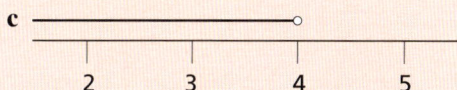

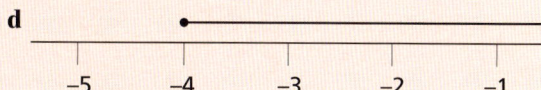

a

b

c

d

2 Make copies of this number line and on them illustrate these inequalities.

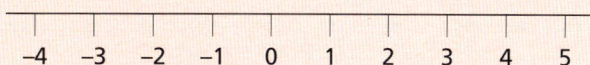

a $x > -2$	**b** $x < 3$	**c** $x \geq -3$	**d** $x \leq 4$
e $x \geq -1$	**f** $x \leq 0$	**g** $x > 2$	**h** $x < -2$

3 Using a number line or otherwise, for each of the following write down the least integer (whole number) which obeys the inequality.

a $x > 3$	**b** $x > -2$	**c** $x \geq -4$	**d** $x \geq 0$
e $x > \frac{1}{2}$	**f** $x \geq -2.3$	**g** $x > 1\frac{1}{3}$	**h** $x \geq \frac{7}{3}$

4 For each of the following write down the greatest integer which obeys the inequality.

a $x < 7$	**b** $x \leq 3$	**c** $x \leq -5$	**d** $x < -7$
e $x < \frac{1}{4}$	**f** $x \leq -1\frac{2}{3}$	**g** $x < -6.2$	**h** $x \leq -3\frac{3}{8}$

Solving inequalities

An inequality is like an equation, but with the = sign replaced by one of the four inequality signs. The rules for solving an inequality are the same as for solving an equation, with one exception.

● *When you multiply or divide an inequality by a negative number, the inequality sign reverses.*

So if $x < y$, then $-x > -y$. If $a \geq b$, then $-a \leq -b$.

Justification

If you have £100 and I have £50, then you are richer than I. But if you owe £100 and I owe £50, then you are poorer than I. Changing £100 to $-£100$ and £50 to $-£50$ also changes the word 'richer' to 'poorer'.

$$100 > 50 \quad \text{but} \quad -100 < -50$$

Examples Solve the inequality $4x - 5 < 2x + 11$.

Subtract $2x$ from both sides, and add 5 to both sides.

> The process involved is identical to that for solving the equation $4x - 5 = 2x + 11$.

$$2x < 16$$

Divide both sides by 2. As 2 is positive, the inequality is unchanged.

$$x < 8$$

Solve the inequality $3 - \frac{1}{2}x \le 1 - \frac{1}{6}x$.

Clear up the fractions, by multiplying both sides by 6. $\frac{1}{2} \times 6 = 3$, and $\frac{1}{6} \times 6 = 1$.

$$18 - 3x \le 6 - x$$

Add x to both sides, and subtract 18 from both sides.

$$-2x \le -12$$

Divide both sides by -2. This is negative, so change the inequality sign from $\le$ to $\ge$.

$$x \ge 6$$

Note. You might find it easier to collect the xs on the right-hand side, so that they become positive.

$$18 \le 6 + 2x$$
$$12 \le 2x$$
$$6 \le x$$

This means the same as $x \ge 6$.

The length of a rectangle is 3 cm greater than the width. If the perimeter is less than 20 cm, find the range of values of the width.

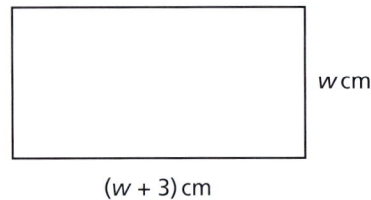

$(w + 3)$ cm

Let the width be w cm. Then the length is $(w + 3)$ cm. The perimeter is the sum of twice the length and twice the width. This is less than 20 cm.

$$2(w + 3) + 2w < 20$$
$$2w + 6 + 2w < 20$$
$$4w < 14$$
$$w < 3.5$$

The width is less than 3.5 cm.

Exercise 9.3

For questions 1–30, solve the inequalities, showing your answers on a number line.

1 $3x - 2 < 4$	**2** $2x + 3 > 5$	**3** $3x - 1 \le 8$
4 $4x + 3 \ge 7$	**5** $5x - 2 > 23$	**6** $3x - 1 \ge 26$
7 $2x + 11 < 5$	**8** $4x + 10 \le 2$	**9** $3x + 1 < x + 11$
10 $4x - 3 \ge x + 3$	**11** $2x + 5 > 5x - 4$	**12** $7x - 3 \le 2x - 17$
13 $3 - 2x \le 7 - 4x$	**14** $1 - 5x > 2 - 2x$	**15** $6 - x \le 3 - 3x$

16 $7 - 2x \geq 3x - 13$ **17** $6x + 2 < 16 - x$ **18** $\frac{1}{2}x + 3 < \frac{1}{3}x + 1$
19 $\frac{1}{4}x + 3 \leq \frac{3}{8}x + 1$ **20** $\frac{2}{3}x - 7 \geq \frac{5}{8}x + 3$ **21** $2 - \frac{1}{3}x \geq 4 - \frac{1}{2}x$
22 $6 - \frac{1}{4}x \geq 1 - \frac{1}{5}x$ **23** $0.3x + 2 > 0.4x - 5$ **24** $0.8x - 5 \geq 0.6x + 7$
25 $5 - 0.7x > 0.1x - 3$ **26** $2(x + 3) \leq 3(x - 7)$ **27** $5(x - 4) > 3(x + 6)$
28 $3(2x - 7) \geq 2(4x - 11)$ **29** $4(3x + 1) < 2(7x - 8)$ **30** $0.3(x + 3) \geq 0.4(x - 5)$

31 The length of a rectangle is 5 m greater than the width. If the perimeter is more than 30 m, find the range of values of the width.

32 The width of a rectangle is 7 cm less than the length. If the perimeter is at least 40 cm, what is the length?

33 Anne has £9 more than Belinda. Together they have less than £50. How much does Belinda have?

34 Aziz has three times as much money as Rashid. Even if Aziz gives Rashid £240, he will still be richer. How much does Rashid have?

35 A fully-laden ship is carrying 2000 economy cars and 1500 luxury cars. Each luxury car weighs 500 kg more than an economy car. The total cargo can weigh at most 5 000 000 kg. What is the weight of an economy car?

Combined inequalities

Two or more inequalities can be combined. If $x > 1$ and $x < 5$, then we can write $1 < x < 5$. This is shown on the number line below.

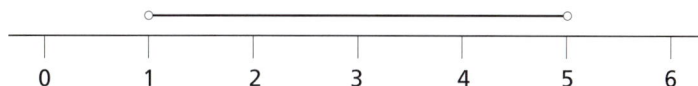

Examples   Illustrate $-1 \leq x < 3$ on the number line. List the integers obeying the inequality.

The line starts with a filled-in dot at -1, then goes up to a hollow dot at 3, as shown.

The integers go from -1 to 3, and -1 is included but 3 is not.
   The integers are $-1, 0, 1, 2$.

Solve the inequality $2 < 3x + 5 < 14$.

Subtract 5 from all three terms,

   $-3 < 3x < 9$

then divide by 3.
   So $-1 < x < 3$.
   The result is shown on this number line.

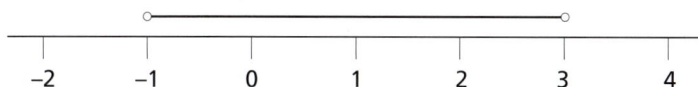

Exercise 9.4

1 Illustrate each of these inequalities on a number line. In each case, list the integers which obey the inequality.

   a $-2 < x \leq 2$ **b** $0 \leq x \leq 4$

   c $-3 < x < 2$ **d** $1 \leq x < 5$

   e $-1 \leq x \leq 3$ **f** $3 \leq x < 5$

   g $-4 < x < 0$ **h** $-5 < x \leq 1$

2 The integers $-1, 0, 1, 2, 3, 4$ obey a combined inequality. Write down four different inequalities it could be.

3 For each of the following sets of integers, write down four different inequalities which it obeys.

   a $2, 3, 4$ **b** $-4, -3, -2, -1$

Solve the inequalities in questions 4–11, illustrating your answers on a number line.

4 $1 < 3x - 5 < 7$ **5** $2 \leq 2x - 6 \leq 8$

6 $5 \leq 4x + 1 < 17$ **7** $-7 < 2x - 3 \leq 7$

8 $-5 \leq 2x + 1 < 3$ **9** $2 < 5x - 8 \leq 22$

10 $3 < 2x + 11 < 5$ **11** $-7 \leq 3x - 4 \leq -4$

Examples   Solve the inequality $x + 7 < 5x + 3 \leq 4x + 11$.

This is the same as the two inequalities, $x + 7 < 5x + 3$ and $5x + 3 \leq 4x + 11$.

   For $x + 7 < 5x + 3$

$$7 < 4x + 3 \quad \text{hence} \quad 4 < 4x$$

This gives $1 < x$.

   For $5x + 3 \leq 4x + 11$

$$x + 3 \leq 11 \quad \text{hence} \quad x \leq 8$$

$1 < x \leq 8$.

Solve the inequality $2 < 3x - 7 \leq 4x + 1$.

As above, solve the two inequalities $2 < 3x - 7$ and $3x - 7 \leq 4x + 1$.

   For $2 < 3x - 7$

$$9 < 3x \quad \text{hence} \quad 3 < x$$

For $3x - 7 \leq 4x + 1$

$$-7 \leq x + 1 \quad \text{hence} \quad -8 \leq x$$

Both these inequalities are shown. We want both inequalities to hold, so we want the region under *both* the bars. This is the region under the first bar.

$$3 < x$$

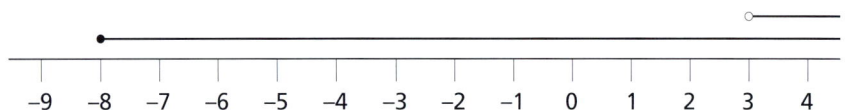

Exercise 9.5

Solve these inequalities, illustrating your answers on a number line.

1 $2 \leq 4x - 6 < 3x + 1$

2 $x + 3 < 2x + 1 < 9$

3 $5x - 3 < 2x + 6 \leq 3x + 8$

4 $2 + x \leq 1 + 2x \leq x + 5$

5 $1 - x < 5 - 3x \leq 1 - 4x$

6 $4 - 2x < 8 + 2x < 3x - 7$

7 $3 - x \leq 5 + x \leq 7 - x$

8 $1 - 3x \leq x < 1 - 2x$

9 $\frac{1}{2}x + 3 < 2x < \frac{1}{3}x + 5$

10 $0.1x + 5 < 0.5x + 1 < 0.4x + 3$

11 $2x + 3 \leq 4x + 1 \leq 3x + 2$

12 $2x + 3 \leq 3x - 1 \leq x + 5$

Two-dimensional inequalities

The inequalities of the previous section involved one variable. You can get inequalities with two variables, such as $x + y \leq 5$. The solution to this sort of inequality consists of lots of *pairs* of values of x and y, such as $x = 2$ and $y = 1$, that is, $(2, 1)$. So the solution consists of a two-dimensional region. We can show this region on a graph.

The line with equation $x + y = 5$ consists of points $(0, 5)$, $(1, 4)$, $(2, 3)$ and so on, which all lie on a straight line as shown in the diagram.

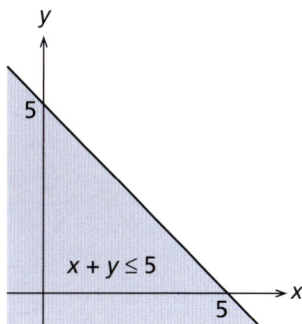

Going above the line will increase the value of y, so this region will contain points for which $x + y > 5$. This is shaded red.

Going below the line will decrease the value of y, so this region will contain points for which $x + y < 5$. This is shaded blue.

The inequality $x + y \leq 5$ includes the line $x + y = 5$. The boundary is shown as a solid line.

If the inequality is $x + y < 5$, then it doesn't include the line $x + y = 5$. The boundary is shown as a broken line. The diagrams below show the two cases.

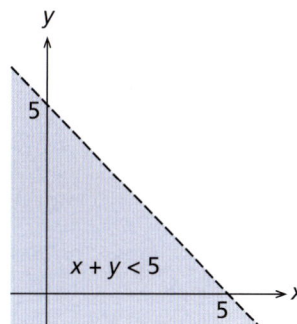

Examples

In the diagram, the straight line has equation $3y + 2x = 6$. What inequality is obeyed in the region shaded blue?

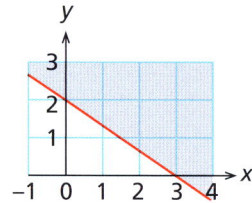

The region is above the line. So y is larger in this region. The value of $3y + 2x$ has increased from 6.

Note that the boundary is a solid line. So the line itself is included.

The inequality is $3y + 2x \geq 6$.

Check: Take a point in the blue region, say $(2, 2)$. The value of $3y + 2x$ at $(2, 2)$ is $3 \times 2 + 2 \times 2$, which is 10. This is greater than 6. We have the correct region.

In the diagram, the straight line has equation $y - 2x = 4$. What inequality is obeyed in the region shaded green?

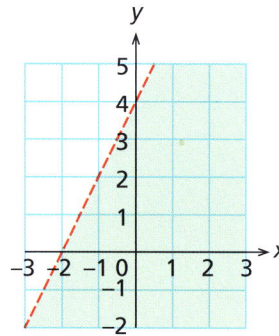

The region is below the line. So y is smaller in this region. The value of $y - 2x$ has decreased from 4.

Note that the line is broken. So the line itself is not included.

The inequality is $y - 2x < 4$.

Check: Take a point in the green region, say $(1, 3)$. The value of $y - 2x$ at $(1, 3)$ is $3 - 2 \times 1$, which is 1. This is less than 4. We have the correct region.

Exercise 9.6

For each of these questions, give the inequality obeyed in the shaded region. The equation of each line is given.

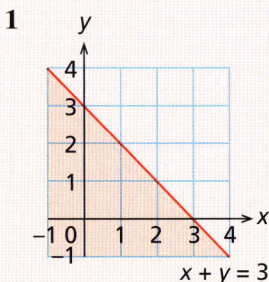

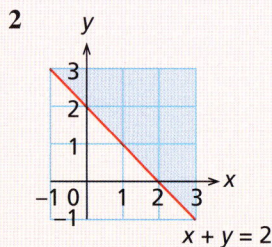

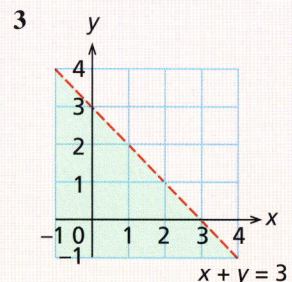

1

$x + y = 3$

2

$x + y = 2$

3

$x + y = 3$

4

5

6 $4y + 3x = 12$

$y - x = 1$

$y - x = -1$

7 $5y + 2x = 10$

8 $2y - 3x = 6$

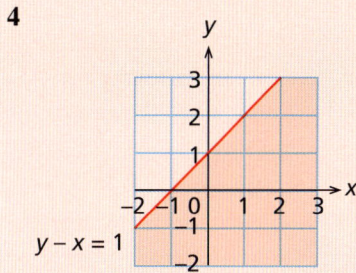

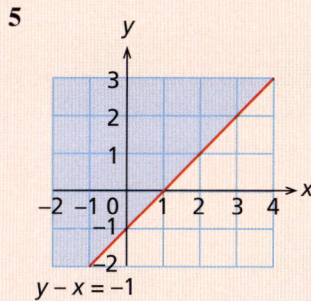

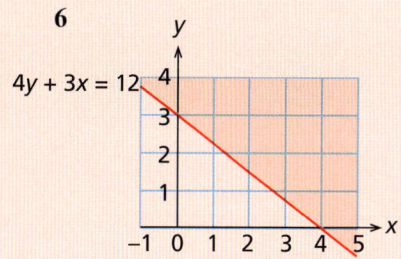

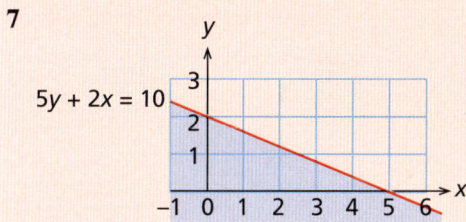

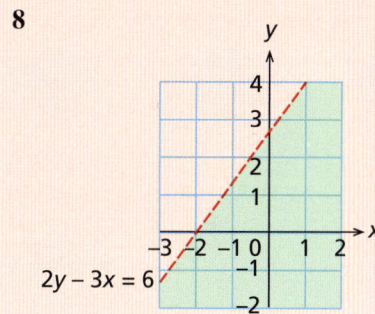

Several regions

A region bounded by more than one line will obey more than one inequality. To avoid having different shadings on top of each other, it is usual to shade the region we *don't* want. The region we *do* want is left unshaded.

Examples The lines in this diagram are $y - 2x = 4$, $y = 1$ and $y + x = 5$. Give the inequalities obeyed in the triangle enclosed in the middle.

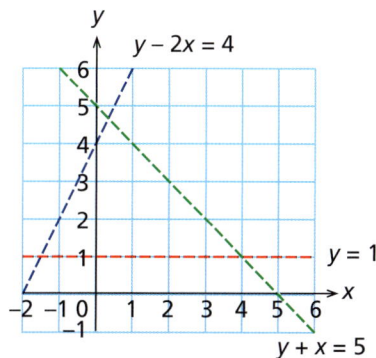

$y - 2x = 4$

$y = 1$

$y + x = 5$

Remember:

It doesn't matter whether you shade in the region you *do want, or shade out the region you* don't *want. But it must be clear to whoever sees your work which representation you have chosen.*

The region is above $y = 1$. Hence $y > 1$.
The region is below $y - 2x = 4$. Hence $y - 2x < 4$.
The region is below $y + x = 5$. Hence $y + x < 5$.

The inequalities are $y > 1$, $y - 2x < 4$ and $y + x < 5$.

Shade the diagram on page 127 to show the region satisfied by the inequalities $y > 1$, $y - 2x > 4$ and $y + x > 5$.

We want the region above the line $y = 1$, above $y - 2x = 4$, and above $y + x = 5$. Shade out the regions below these lines with red, blue and green. The region we want is the top unshaded region.

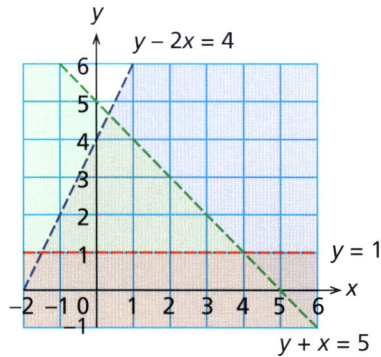

Check: A point in this region is (1, 7).
　　Certainly $y > 1$.

　　　$y - 2x = 7 - 2 \times 1$, which is 5. This is greater than 4.
　　　$x + y = 8$, which is greater than 5.

We have the correct region.

Exercise 9.7

The diagram shows regions bounded by the lines $x = 1$, $2x + 3y = 6$ and $y - 3x = 0$.

You will need:
- graph paper
- colouring pens/pencils

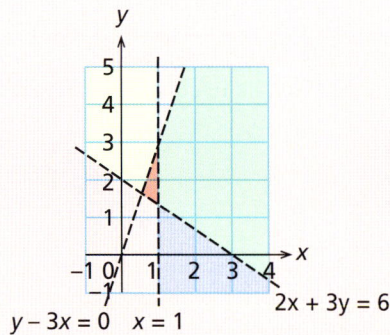

For questions 1–4, give the inequalities obeyed in the region shaded:

1 red　　　　　　　　**2** blue　　　　　　　　**3** yellow　　　　　　　　**4** green.

For each of questions 5–7, make a copy of the axes and lines of the diagram above, and indicate which region satisfies the given inequalities. Shade out the regions you don't want.

5 $x < 1$, $2x + 3y < 6$ and $y - 3x > 0$
6 $x > 1$, $2x + 3y > 6$ and $y - 3x > 0$
7 $x < 1$, $2x + 3y < 6$ and $y - 3x < 0$

The diagram shows regions bounded by the lines $y = -1$, $5x + 2y = 10$ and $y - 3x = 3$.

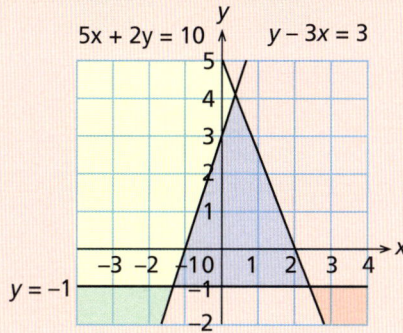

For questions 8–11, give the inequalities obeyed in the region shaded:

8 red **9** blue **10** yellow **11** green.

For each of questions 12–14, make a copy of the axes and lines of the diagram above, and indicate which region satisfies the given inequalities. Shade out the regions you don't want.

12 $y \leq -1$, $5x + 2y \leq 10$ and $y - 3x \leq 3$
13 $y \geq -1$, $5x + 2y \geq 10$ and $y - 3x \geq 3$
14 $y \geq -1$, $5x + 2y \geq 10$ and $y - 3x \leq 3$

Illustrating two-dimensional inequalities

To find the region corresponding to a two-dimensional inequality, first draw the line corresponding to the equation. Then decide which side of the line you want, and shade it.

Examples   Illustrate the inequality $x + y < 3$.

First draw the line $x + y = 3$. This line goes through $(0, 3)$, $(1, 2)$, $(2, 1)$ and so on. The inequality is $<$, so the line itself is not included. Draw a broken line in the diagram.

We want $x + y$ to be less than 3. So we want the region below the line. This is shaded red in the diagram.

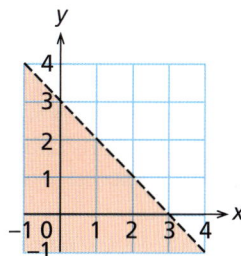

I buy x cakes at 40p each, and y scones at 30p each. I spend at most £2.40. Write down an inequality in x and y. Illustrate this inequality.

The cost of the cakes is $40x$ pence, and the cost of the scones is $30y$ pence. The total cost is less than £2.40, that is, 240p.

> There are £ and p involved here!

$$40x + 30y \leq 240$$

This inequality simplifies to $4x + 3y \leq 24$. The line $4x + 3y = 24$ goes through $(0, 8)$ and $(6, 0)$. The inequality is $\leq$, so the line itself is included. Draw a solid line on a graph, and shade the region below it. The result is shown in the diagram.

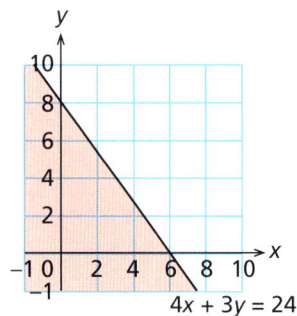

$4x + 3y = 24$

Note. Obviously the numbers cannot be negative, so there are two more inequalities, $x \geq 0$ and $y \geq 0$. The graph only shows values greater than or equal to 0.

Exercise 9.8

For questions 1–22, illustrate the given inequalities on graph paper.

You will need:
- graph paper
- colouring pens/pencils

1 $x > 1$	**2** $y \leq 2$	**3** $x \geq -1$	
4 $y > -2$	**5** $x + y < 6$	**6** $x + y \geq 3$	
7 $2x + 3y \leq 6$	**8** $3x + y \geq 6$	**9** $5x + 2y < 10$	**10** $3y + 5x \leq 15$
11 $y + 3x \geq 9$	**12** $y - x \leq 0$	**13** $y - x < 3$	**14** $y - 2x < 3$
15 $y - 3x \geq 6$	**16** $2y - x \leq 4$	**17** $2y - x > 6$	**18** $3y - x \geq 9$
19 $3y - x \leq 6$	**20** $2y - 3x \leq 6$	**21** $5y - 2x > 10$	**22** $3y - 5x \leq 15$

23 Illustrate the region corresponding to the three inequalities $y \geq 1$, $x + y < 5$ and $y - 2x < 0$.

24 Illustrate the region corresponding to the three inequalities $x \geq -1$, $x + y < 3$ and $2y - x \leq 4$.

25 Illustrate the region corresponding to the three inequalities $y \geq 0$, $2x + 3y \leq 6$ and $3y - 2x < 6$.

26 A ship is carrying x cars at 1 tonne each, and y lorries at 3 tonnes each. The total weight is less than 27 tonnes. Write down an inequality in x and y. Illustrate it on graph paper.

27 A firm orders 10 computers at £x each, and 3 printers at £y each. The total bill is at most £10 000. Write down an inequality in x and y. Illustrate it on graph paper.

28 A man works x hours normal time at £8 per hour, and y hours overtime at £12 per hour. He earns more than £380. Write down an inequality in x and y. Illustrate it on graph paper.

29 On an electrical circuit there are x light bulbs at 100 watts each, and y heaters at 1000 watts each. The total load on the circuit must not exceed 5000 watts. Write down an inequality in x and y and illustrate it on graph paper.

30 In an amusement arcade, *Kendo Kombat* costs xp per go, and *Wing Chun Warrior* costs yp. I have two goes on *Kendo Kombat* and one on *Wing Chun Warrior*, spending at most £4. Write down an inequality in x and y. Illustrate it on graph paper.

Linear programming

In many situations there is some value we want to make as large as possible or as small as possible. For example

A business wants to make the profits as large as possible.
A manufacturer wants to make the costs as small as possible.

The value that we want to make as large or as small as possible is the **objective function**. If all the constraints and the object function are linear, then the process of finding the greatest or least value of the objective function is called **linear programming**. This greatest or least value is at one of the corners of the region of possible solutions. The method of finding the greatest or least value is shown in the next example.

Example   A theatre has a maximum number of 400 seats. Each cheap seat occupies $1 \, m^2$ of space, and each expensive seat occupies $2 \, m^2$. There is $600 \, m^2$ of space available.

Suppose x cheap and y expensive seats are allocated. Write down inequalities in x and y and illustrate them on a graph.

If a cheap seat costs £8 and an expensive seat costs £12, what allocation of seats will bring in the most money?

The constraints are

$$x + y \leq 400 \quad \text{(the constraint for total number)}$$
$$x + 2y \leq 600 \quad \text{(the constraint for space)}$$

Obviously, $x \geq 0$ and $y \geq 0$.
The four inequalities are
shown on the diagram.

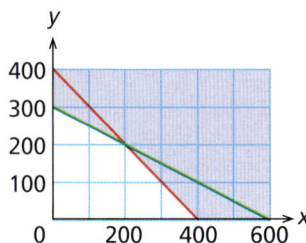

The profit is £$(8x + 12y)$.

On the graph draw a line parallel to $8x + 12y = 0$, that is, with a slope of $-\frac{2}{3}$. It is easier to take a line which crosses the axes at positive integer values: the line $8x + 12y = 2400$ goes through $(300, 0)$ and $(0, 200)$.

Put a ruler along this line, and gradually slide it upwards until it is just about to leave the region. At this point the maximum value is found.

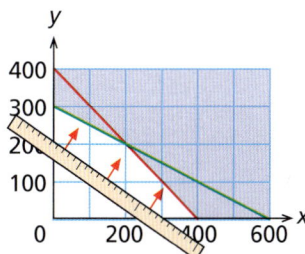

The point is $(200, 200)$. The value of the objective function is

$$8 \times 200 + 12 \times 200 = 4000$$

The maximum revenue is £4000.

Exercise 9.9

1 Illustrate the inequalities $x \geq 0$, $y \geq 0$, $2x + 3y \leq 12$ and $3x + 2y \leq 12$. Find the greatest value of $x + y$ subject to these inequalities.

2 Illustrate the inequalities $x \geq 0$, $y \geq 0$, $y \geq x$ and $x + 2y \leq 12$. Find the greatest value of $x + 3y$ subject to these inequalities.

3 Illustrate the inequalities $x \geq 0$, $y \geq 0$, $2x + 3y \geq 6$ and $2x + y \geq 4$. Find the least value of $x + y$ subject to these inequalities.

4 A firm makes curtains which are either ordinary or de luxe. Each ordinary curtain takes 3 hours to produce and uses 6 m of material, and each de luxe curtain takes 6 hours to produce and uses 7 m of material. The workers of the firm can work for a total of 60 hours, and there is 90 m of material available. If x ordinary and y de luxe curtains are made, write down inequalities in x and y. Illustrate these inequalities on a graph.

  If the profits on ordinary and de luxe curtains are £6 and £9 respectively, find how many of each should be made to maximise profit.

5 A shop has space for 5 refrigerators. The shop buys them from the manufacturer at £400 for type A, and £500 for type B. There is £2400 available to spend. If x of type A and y of type B are bought, write down inequalities in x and y to illustrate them.

  The profits are £90 for type A refrigerators and £140 for type B. How many of each should be stocked to bring in the maximum profit? What is the maximum profit?

6 A factory manager can buy two sorts of machine: type A uses $4\,m^2$ of space and costs £2000, and type B uses $3\,m^2$ of space and costs £3000. There is $28\,m^2$ of space available and £20 000 to spend. If x type A machines and y type B machines are bought, write down inequalities in x and y and illustrate them.

  a What are the numbers of type A and type B machines which maximise the total number of machines?

  b Suppose that type A produces 40 items per hour, and type B 100 items per hour. What are the numbers of type A and type B which maximise the total production?

7 A factory needs to employ new workers, who are either skilled or unskilled. There must be at most 100 workers, and there must be at least 40 skilled workers.

  Each skilled worker is paid £150 per day, and each unskilled worker is paid £60 per day. The total wages cannot be more than £12 300. Suppose there are x skilled workers and y unskilled workers.

  a Write down inequalities in x and y for the above conditions. Illustrate these inequalities.

  b Each skilled worker produces 60 items per day, and each unskilled worker produces 20 items. Use the graph to find how many workers of each type should be hired to maximise production.

8 A farmer needs to harvest his crops in a week. He can hire two types of machine. Each type A machine will harvest 6 hectares and each type B will harvest 10 hectares. Each type A requires 3 men to operate it, and each type B requires 2 men. There are 60 hectares to be harvested, and there are 18 men available. Assume he hires x of type A and y of type B.

  a Write down inequalities for the above restrictions, and illustrate them on a graph.

  b The cost of hiring the machines is £1000 for type A, and £1600 for type B. How many of type A and type B should be hired to minimise the cost? What is the minimum cost?

9 Food A contains 4 units of protein and 5 units of starch per kg, and food B contains 6 units of protein and 3 units of starch per kg. The minimum daily intake of protein is 16 units, and the minimum daily intake of starch is 11 units.

  Suppose the daily intake is x kg of food A and y kg of food B. Find inequalities in x and y and illustrate them.

  Suppose A costs £8 per kg, and B costs £10 per kg. What are the amounts of A and B which make the total cost as small as possible?

10 An aircraft has $600\,m^2$ of cabin space and can carry $5000\,kg$ of luggage. An economy class passenger gets $3\,m^2$ of space, and is allowed $20\,kg$ of luggage. A first class passenger gets $4\,m^2$ of space and is allowed $50\,kg$ of luggage. There must be space for at least 50 economy class passengers. Let x be the number of economy seats, and y the number of first class seats. Find the inequalities in x and y and illustrate them.

   The profit per flight from an economy seat is £80, and from a first class seat the profit is £160. What allocation of seats will give the maximum profit?

SUMMARY

- The four **inequality** signs are $<$, $>$, $\leq$ and $\geq$.
- Inequalities can be shown by a bar on a **number line**. If the inequality involves $<$ or $>$, the bar ends with a hollow dot. If the inequality involves $\leq$ or $\geq$, then it ends with a solid dot. For example, the bar to illustrate $2 < x \leq 7$ will have a hollow dot at the left and a solid dot at the right.
- An inequality in two variables can be shown by a region on a plane. To illustrate an inequality, first draw the line of the corresponding equation. If the inequality is $\leq$ or $\geq$, draw a solid line. If the inequality is $<$ or $>$, draw a broken line. Then shade one side of this line.
- To find the maximum or minimum value of a function subject to constraints, draw the region corresponding to the constraints, then find the extreme value of the function within the region.

Exercise 9A

You will need:
- graph paper
- colouring pens/pencils

1 Solve the inequality $3x + 7 < 4$.
2 Solve the inequality $2x + 1 \leq 11 - \frac{1}{2}x$.
3 Write down the inequality shown on this number line.

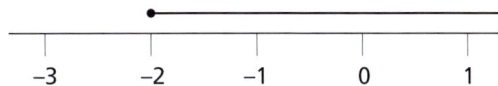

4 Solve the inequality $-3 \leq 4x + 1 < 9$, illustrating your answer on a number line.
5 Solve the inequality $1 + 3x < 4 - x \leq 2x + 8$.
6 The line in the diagram has the equation $3y + 4x = 12$. Write down an inequality obeyed in the region shaded red.

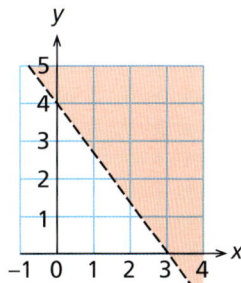

7 The lines in the diagram have equations $x = -1$, $y = 1$ and $x + y = 6$. Write down the inequalities satisfied by the triangular region in the middle of the diagram.

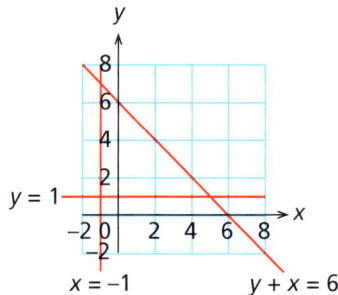

8 Make a rough copy of the diagram of question 7, and on it indicate the region which satisfies the inequalities $x > -1$, $y < 1$ and $x + y < 6$. Shade out the regions you don't want.

9 Vesuvius pizzas cost £4 each and weigh 500 g, and Stromboli pizzas cost £3 each and weigh 625 g. I buy x Vesuvius pizzas and y Stromboli pizzas. I spend at most £20 and the total weight is at most 3500 g. Write down inequalities in x and y.

10 On graph paper illustrate the inequalities of question 9. Find the greatest number of pizzas I could have bought.

Exercise 9B

You will need:
• graph paper
• colouring pens/pencils

1 Solve the inequality $3 > 1 - 2x$.

2 Solve the inequality $\frac{1}{6}x + 3 \leq \frac{1}{4}x - 7$.

3 Write down the inequality shown on this number line.

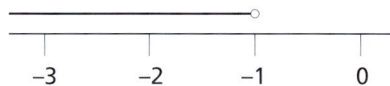

4 Solve the inequality $-5 < 4x + 7 \leq 19$, illustrating your answer on a number line.

5 Solve the inequality $\frac{1}{2}x - 1 \leq \frac{1}{3}x + 2 < \frac{1}{4}x + 7$, illustrating your answer on a number line.

6 The line in the diagram has equation $2y - 3x = 1$. Write down an inequality obeyed by the region shaded blue.

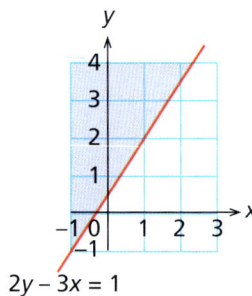

7 The lines in the diagram have equations $y = 2$, $y = x$ and $y + x = 6$. Write down the inequalities obeyed by the region shaded green.

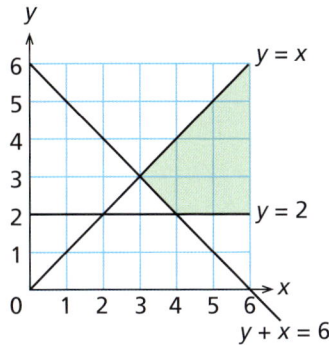

8 Make a rough copy of the diagram of question 7, and on it indicate the region which obeys the inequalities $y > 2$, $y > x$ and $y + x < 6$. Shade out the regions you don't want.

9 At a theme park, the Deathfall ride costs £1.50 and the Nightmare ride costs £2. I have x goes on the Deathfall and y goes on the Nightmare. I spent £10 at most and I had at most 6 rides. Write down inequalities in x and y.

10 Illustrate the inequalities of question 9. If the Deathfall lasts 5 minutes and the Nightmare 6 minutes, what is the greatest length of time I could spend on the rides?

Exercise 9C (Ma1)

Which of the following statements are always true? Give a justification for each true one, and a counterexample for each false one.

1 If $x < y$ and $a < b$, then $x + a < y + b$.
2 If $x < y$ and $a < b$, then $x - a < y - b$.
3 If $x < y$ and $a < b$, then $x \times a < y \times b$.
4 If $x < y$ and $a < b$, then $x \div a < y \div b$.
5 If $x < y$ and $y < z$, then $x < z$.

Exercise 9D (Ma1)

Sometimes compound inequalities can be simplified. In the example on page 124, the compound inequality $3 < x$ and $-8 \leq x$ was simplified to $3 < x$.
Simplify these inequalities where possible.

1 $x < 3$ and $x < 7$
2 $x > 5$ and $x \geq 2$
3 $x < 2$ and $x \geq 4$
4 $x < 3$ or $x < 1$
5 $x \geq 3$ or $x \geq 1$
6 $x < 1$ or $x > 0$
7 $(x > 1$ or $x > 3)$ and $(x > 1$ or $x > 0)$
8 $(x > 3$ and $x < 6)$ or $(x < 4$ and $x > 0)$

> In mathematics the word 'or' is always inclusive. That is, 'A or B' includes A and B.

Exercise 9E

All the inequalities in this chapter have been linear. Inequalities involving x^2, $\frac{1}{x}$, and so on, are more complicated.

Examples of these inequalities are below. In many cases it is best to draw a graph.

Solve these inequalities.

1 $x^2 \leq 9$

2 $\frac{1}{x} < 2$

3 $\frac{1}{x} \geq -3$

4 $x^2 - 5x + 6 < 0$ (Hint: factorise.)

5 $x^2 + 4x - 12 \geq 0$

6 $x^2 - 4x + 2 < 0$

7 $\cos x \geq 0.5$

8 $\sin x > 0.3$

10 Graphs

Types of graph

Already you have met many types of **graph**. Here are some common ones. An equation of the form $y = mx + c$ (m and c constant) has a straight line or **linear** graph. To draw this graph, plot two pairs (x, y) which obey the equation, and join them with a straight line.

An equation of the form $y = ax^2 + bx + c$ has a **quadratic** graph. It will look like either of the shapes below. The curve is called a **parabola**.

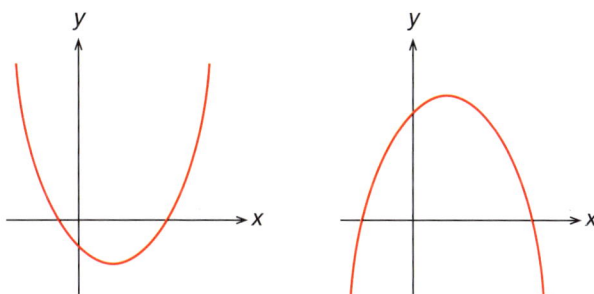

If $a > 0$, then the shape is like the graph on the left. This has a lowest point, called a **minimum**.

If $a < 0$, then the shape is like the graph on the right. This has a highest point, called a **maximum**.

You can draw a rough sketch of the graph if you know where it crosses the axes.

An equation of the form $y = \dfrac{a}{x - b}$ has a **reciprocal** graph. It will look like the shape below. Notice that there is no point on the graph for $x = b$: if you put $x = b$, the value of y is $\dfrac{a}{b - b} = \dfrac{a}{0}$. This does not exist, as you cannot divide by 0.

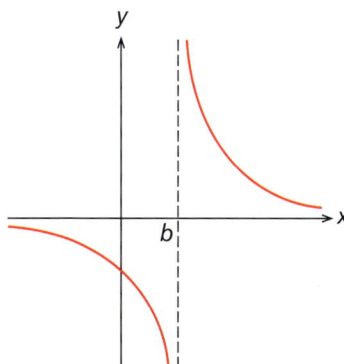

Graphics calculators

A graphics calculator makes it much easier to draw graphs. In particular, you can use it to draw many similar graphs and spot a pattern. The exercises in this chapter contain suggestions for the use of a graphics calculator. They are indicated by a 🖩 in front of the question number.

Examples

Let $y = x^2 - 3x + 2$. Find where $y = 0$ and where $x = 0$, and hence sketch the graph of $y = x^2 - 3x + 2$.

If $y = 0$, then $x^2 - 3x + 2 = 0$. This factorises, to

Remember chapter 4.

$$(x - 1)(x - 2) = 0$$

Hence $x = 1$ or $x = 2$. The graph must go through $(1, 0)$ and $(2, 0)$.
  When $x = 0$, then $y = 2$, so the graph goes through $(0, 2)$.
  The coefficient of x^2 is 1, which is positive, hence the graph has a minimum.
  The sketch graph is shown.

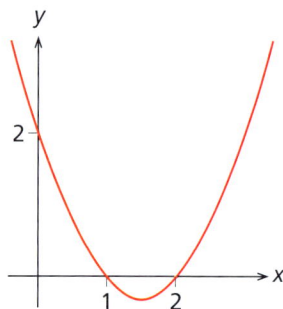

The graph below is of the form $y = \dfrac{1}{x - b}$. Find b.

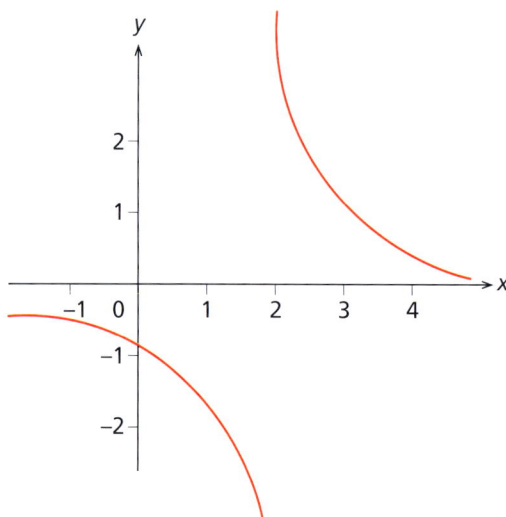

Notice that the graph does not exist when $x = 2$. So $x - b$ must be 0 when $x = 2$. This gives $b = 2$.

Exercise 10.1

1 Identify the graphs below as linear, quadratic or reciprocal.

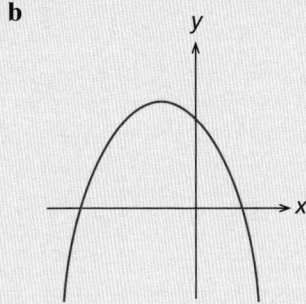

a

b

c

2 The graphs below are of the form $y = mx + c$. In each case find the values of m and c.

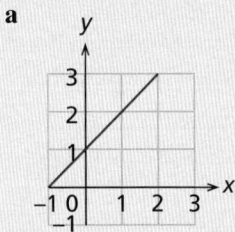

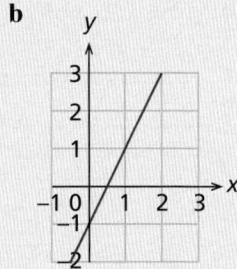

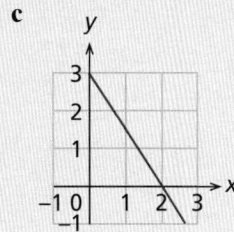

a

b

c

3 The graphs below are of the form $y = \dfrac{1}{x - b}$. In each case find b.

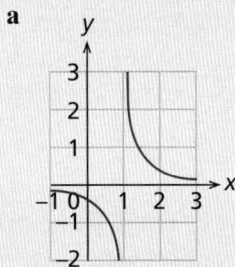

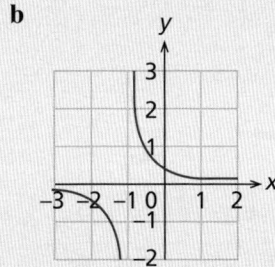

a

b

4 The graphs below are of the form $y = x^2 - bx + c$. In each case find b and c.

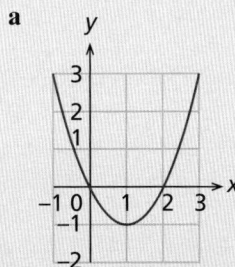

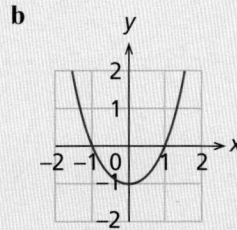

a

b

5 Sketch the graph of $y = x^2 - 5x + 6$.
6 Sketch the graph of $y = x^2 - 2x - 3$.
7 Sketch the graph of $y = x^2 + 4x + 5$.
8 Sketch the graph of $y = -x^2 + 4x - 3$.
9 Sketch the graph of $y = -2x^2 + 5x - 2$.

10 Sketch the graph of $y = -3x^2 + x + 4$.

11 Sketch the graph of $y = \dfrac{2}{x-1}$.

12 Sketch the graph of $y = \dfrac{1}{x-3}$.

13 Sketch the graph of $y = \dfrac{3}{x+2}$.

14 Use a graphics calculator to sketch the following graphs. What do they have in common? Can you generalise?

$$y = x^2 + 2x - 1 \qquad y = x^2 + 2x \qquad y = x^2 + 2x + 1 \qquad y = x^2 + 2x + 2$$

Plotting graphs

Suppose an equation gives y as a function of x. To plot a graph of the function, set up a table of values of x and y, and plot the points. Join up with smooth curves. Be careful if the function does not exist for a particular value of x.

Examples Plot the graph of $y = x^3 + 3x^2 - 1$. Take the values of x between -3 and 1.

Fill in the table below. This gives us:

x	−3	−2	−1	0	1
x^3	−27	−8	−1	0	1
$3x^2$	27	12	3	0	3
−1	−1	−1	−1	−1	−1
y	−1	3	1	−1	3

Plot points at $(-3, -1)$, $(-2, 3)$ and so on on a graph and join with a smooth curve. The result is as shown.

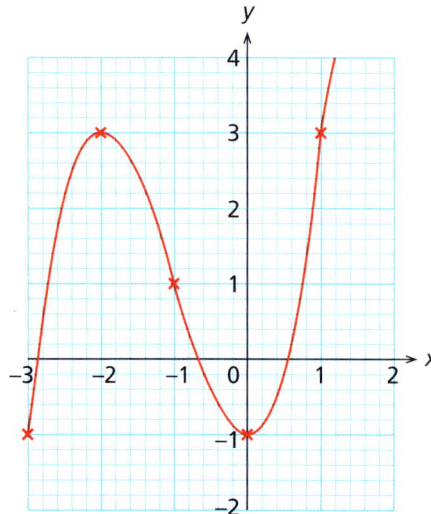

Plot the graph of $y = x + \dfrac{1}{x^2}$, taking values of x between -2 and 2, at intervals of 0.5.

Notice that y is not defined for $x = 0$. So the table must not include it.

x	-2	-1.5	-1	-0.5	0.5	1	1.5	2
$\dfrac{1}{x^2}$	0.25	0.44	1	4	4	1	0.44	0.25
y	-1.75	-1.06	0	3.5	4.5	2	1.94	2.25

The graph is as shown. Note that it reaches up the y-axis, but never crosses it. Curves like this are said to be **asymptotic** to the y-axis.

Exercise 10.2

For questions 1–9, plot the graphs of the functions using the values of x given.

1 $y = x^3 + 2x$ $x = -2, -1, 0, 1, 2$
2 $y = x^3 + x^2$ $x = -2, -1, 0, 1, 2$
3 $y = x^3 - 3x + 1$ $x = -2, -1, 0, 1, 2$
4 $y = x^3 - 2x^2 - 1$ $x = -2, -1, 0, 1, 2, 3$
5 $y = x^2 + \dfrac{1}{x}$ $x = -2, -1, -0.5, 0.5, 1, 2$
6 $y = x - \dfrac{1}{x^2}$ $x = -2, -1, -0.5, 0.5, 1, 2$

7 $y = x^2 - \dfrac{1}{x^2}$ $\qquad x = -2, -1, -0.5, 0.5, 1, 2$

8 $y = x^2 - \dfrac{1}{x}$ $\qquad x = -2, -1, -0.5, 0.5, 1, 2$

9 $y = x^2 + \dfrac{1}{x^2}$ $\qquad x = -2, -1, -0.5, 0.5, 1, 2$

10 Use a graphics calculator to draw the graphs of the expressions below. When do they differ in shape, and when are they the same?

$$y = x^3 + x \qquad y = x^3 + 2x \qquad y = x^3 - x \qquad y = x^3 - 2x \qquad y = x^3 + 0.5x$$

Remember:

We solve simultaneous equations by finding where the graphs of the equations cross.

Intersections of graphs

Suppose two graphs cross. Then, where they cross, both their corresponding equations are true. Two expressions in x are equal to each other. This enables us to find the solution to equations.

Examples   The diagram shows the graph of $y = x^3$. By drawing the line $y = x - 2$, solve the equation

$$x^3 - x + 2 = 0$$

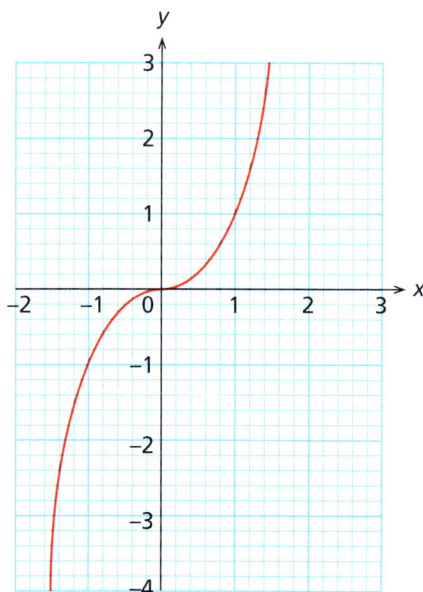

The line $y = x - 2$ goes through $(0, -2)$ and $(2, 0)$. Plot these points, and join them with a straight line.

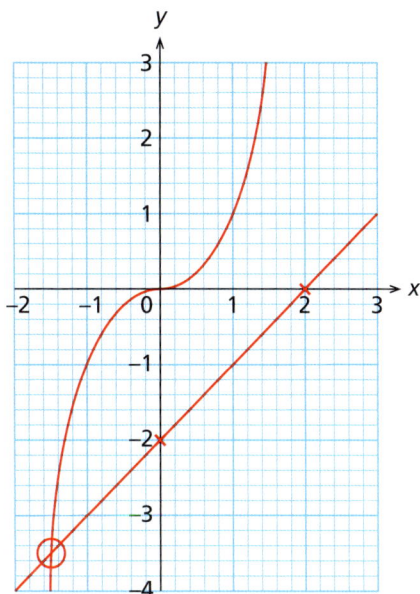

At the crossing point (circled in red) both equations are true. Hence the two expressions in x are equal.

$$y = x^3 = x - 2 \quad \text{hence} \quad x^3 - x + 2 = 0$$

The crossing point is at $x = -1.5$.
 The solution is $x = -1.5$.

Note. We did not give the y-coordinate of the point. The equation we were asked to solve, $x^3 - x + 2 = 0$, only has x in it.
 Check: we can check the answers by testing values on either side of $x = -1.5$.

$$\text{For } x = -1.6, \, x^3 - x + 2 = (-1.6)^3 - (-1.6) + 2 = -0.496 < 0$$
$$\text{For } x = -1.4, \, x^3 - x + 2 = (-1.4)^3 - (-1.4) + 2 = 0.656 > 0$$

So the solution does lie between -1.4 and -1.6.

The diagram shows the graph of $y = x^2$. By drawing a straight line on the graph, solve the equation $x^2 + 3x - 1 = 0$.

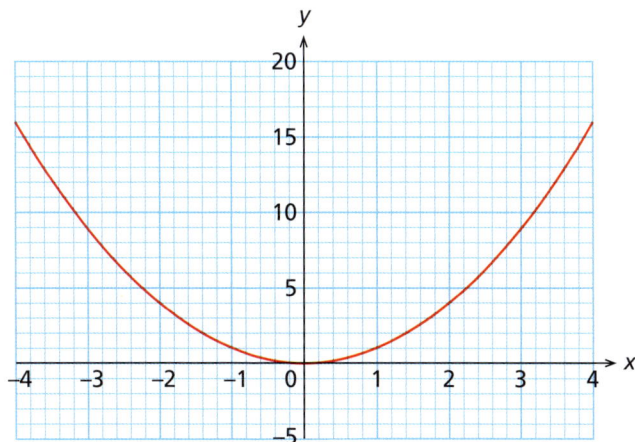

The equation can be arranged as $x^2 = -3x + 1$. We already have the graph of $y = x^2$. If we draw the graph of $y = -3x + 1$, then when they cross

$$x^2 = -3x + 1 \quad \text{hence} \quad x^2 + 3x - 1 = 0$$

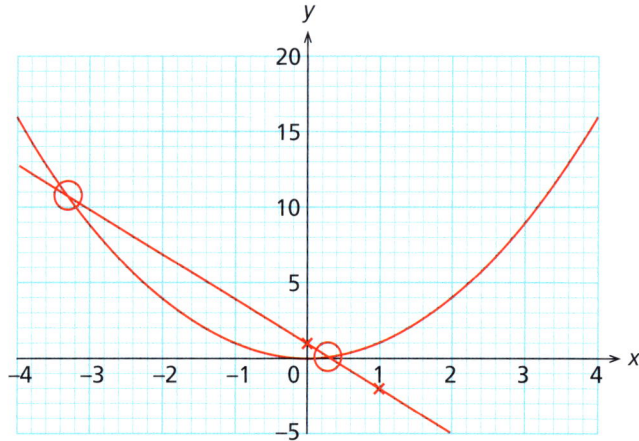

The graph of $y = -3x + 1$ goes through $(0, 1)$ and $(1, -2)$. Plot these points and join them with a ruler. Notice that the line crosses the curve at the points circled in red. The x-coordinates of these points are 0.3 and -3.3.
The solutions are $x = 0.3$ and $x = -3.3$.

Exercise 10.3

You will need:
- graph paper

1 Make a copy of the graph of $y = x^3$ shown.

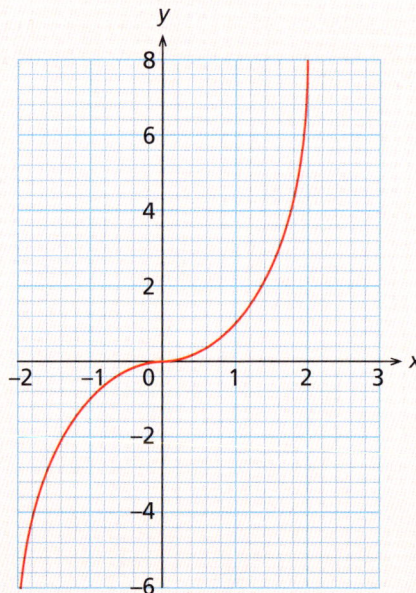

a By drawing the graph of $y = 3 - 2x$, solve the equation $x^3 + 2x - 3 = 0$.
b By drawing the graph of $y = 3x + 1$, solve the equation $x^3 - 3x - 1 = 0$.

c By drawing the graph of $y = \dfrac{1}{x}$, solve the equation $x^4 - 1 = 0$.

2 Make a copy of the graph of $y = x^2$ from the second example on page 143. Use it to solve these equations.

a $x^2 - 2x - 1 = 0$ **b** $x^2 + 3x - 2 = 0$
c $x^2 + x - 3 = 0$ **d** $x^2 - x - 4 = 0$
e $x^2 + 3x + 1 = 0$ **f** $x^2 - 4x + 1 = 0$

g $2x^2 + x - 4 = 0$ **h** $x = \dfrac{1}{x} + 1$

3 Make a copy of the graph on the right of $y = x^2 + 2x - 1$. Use it to solve these equations.

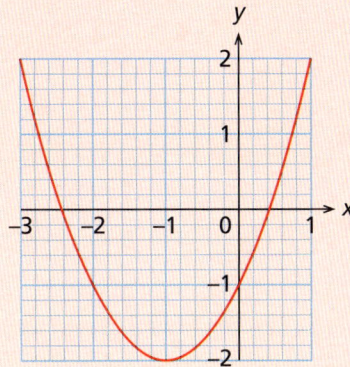

a $x^2 + 2x - 1 = 0$ **b** $x^2 + 2x = 4$
c $x^2 + 3x - 1 = 0$ **d** $x^2 + 3x - 3 = 0$

4 Use a graphics calculator to draw the graph of $y = x^3 - 3x$. Superimpose horizontal lines, of the form $y = k$. Each line crosses the graph at the solutions of the equation $x^3 - 3x = k$.
There are either three solutions, two solutions or one solution. Which values of k give which numbers of solutions?

Exponential increase and decrease

Suppose £1000 is invested at 5% compound interest. Then every year it increases by 5%, that is, it is multiplied by 1.05. After t years it has been multiplied by 1.05 t times, so it is multiplied by 1.05^t. The amount of money is £1000×1.05^t. The graph of the amount is shown below.

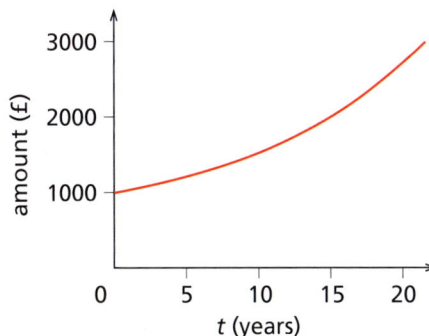

Similarly, suppose 2 kg of radioactive material is decaying at 5% each year. Then every year the amount loses 5%, so it is multiplied by 0.95. After t years it has been multiplied by 0.95 t times, so it is multiplied by 0.95^t. The amount left is 2×0.95^t. The graph of the amount is shown below.

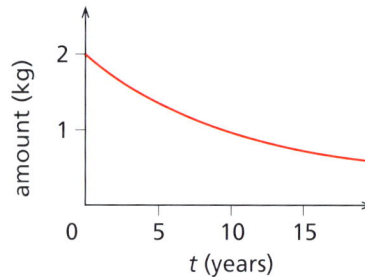

In both these cases the quantity increases or decreases at a rate which is proportional to the quantity itself. That is, the greater the quantity, the greater the increase or decrease. Two other examples are

> Population. The larger a population is, generally the faster it will increase.
> Depreciation. The more expensive a new car is, the faster it will lose its value.

A quantity increasing at a rate proportional to itself is **increasing exponentially**. A quantity decreasing at a rate proportional to itself is **decreasing exponentially**.

If y is increasing or decreasing exponentially over time, then its value at time t is given by

$$y = ab^t \quad \text{where } a \text{ and } b \text{ are constant}$$

If $b > 1$ then y is increasing, and if $0 < b < 1$ then y is decreasing.

> For the money example on page 145, $a = 1000$ and $b = 1.05$. Note that $b > 1$.
> For the radioactivity example above, $a = 2$ and $b = 0.95$. Note that $0 < b < 1$.

In a power like b^t, another word for the index t is **exponent**. This is where the word **exponential** comes from.

You can find the values of a and b from pairs of values of y and t. In particular, if we put $t = 0$, then $b^0 = 1$. Hence $y = a$. So a is the value of y when $t = 0$, that is, it is the initial value of y.

Examples   This graph shows a quantity y which is increasing exponentially over time. Find the equation giving y in terms of t.

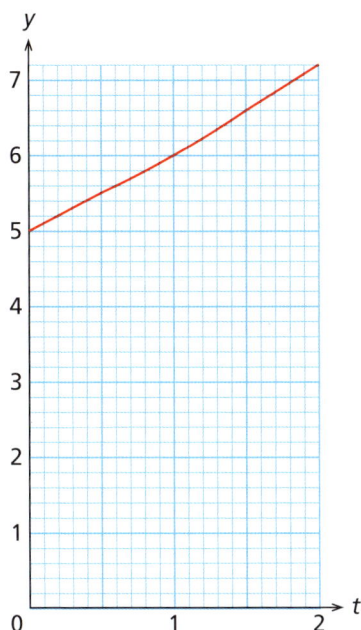

The graph goes through (0, 5) and (2, 7.2). Put these values into the equation $y = ab^t$.

$$5 = ab^0 \qquad 7.2 = ab^2$$

As $b^0 = 1$, the first equation immediately gives $a = 5$. Put this into the second equation.

$$7.2 = 5b^2$$
$$1.44 = b^2$$

> *y is positive, so we ignore the negative square root.*

Hence $b = \sqrt{1.44} = 1.2$.
   The equation is $y = 5 \times 1.2^t$.

For the example above, find y when $t = 3.5$.

Substitute $t = 3.5$ into the equation.

$$y = 5 \times 1.2^{3.5} = 9.46$$

The value of y is 9.46.

Exercise 10.4

1 £2000 is invested at 8%. How much is there after t years?

2 A population is increasing at 3%. If it starts at 15 000 000, how many people are there after t years?

3 10 kg of radioactive material is decreasing at 15% each year. How much is left after t years?

4 A computer is bought for £1600, but its value decreases at 40% each year. How much is it worth after t years?

5 In the graph shown, y is increasing exponentially over time. Find the values of y when $t = 0$ and $t = 1$. Hence find the equation giving y in terms of t.

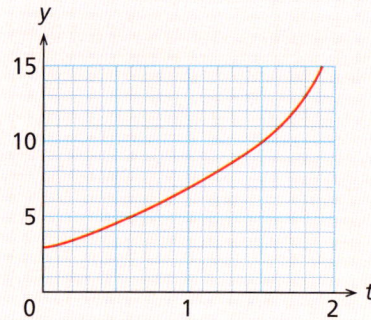

6 In the graph shown, y is decreasing exponentially over time. Find the values of y when $t = 0$ and $t = 2$. Hence find the equation giving y in terms of t.

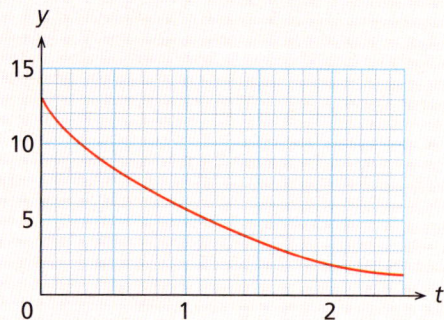

7 A population is increasing exponentially over time, as shown in the graph. Obtain the equation of the graph in the form $P = ab^t$. According to this function, what will the population be in 25 years' time?

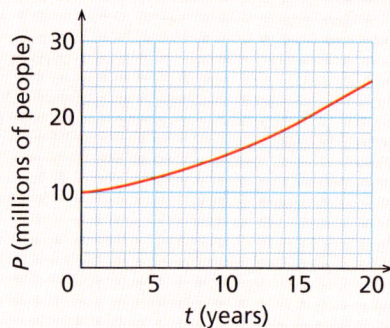

8 A mould is growing exponentially over time, as shown in the graph. Obtain the equation of the graph in the form $M = ab^t$. According to this function, what will the mass of mould be in 3 hours' time?

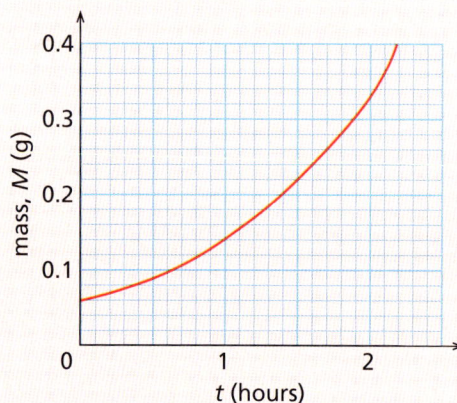

9 The value of a car is decreasing exponentially over time, as shown in the graph. Obtain the equation of the graph in the form $V = ab^t$. According to this function, what will the car be worth after 10 years?

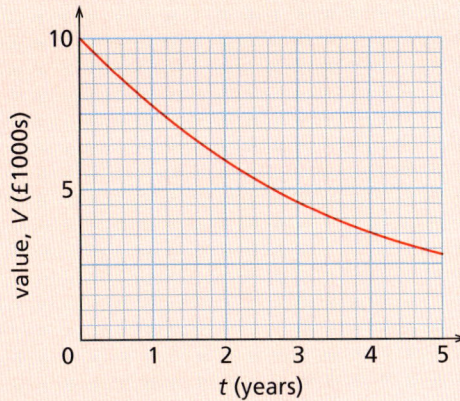

10 The Ruritanian currency, the crown, is depreciating. Its value in terms of dollars is decreasing exponentially, as shown in the graph. Find the equation of the graph, and predict the value of the crown after 35 years.

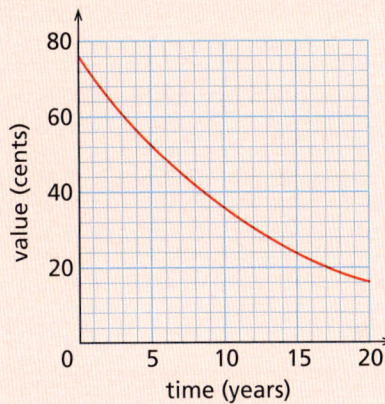

Other quantities

A graph of change over time can show many things. If the quantity is changing at a steady rate, then its graph is straight. If it is changing at an increasing rate then its graph is curved upwards, and if it is changing at a decreasing rate its graph is curved downwards.

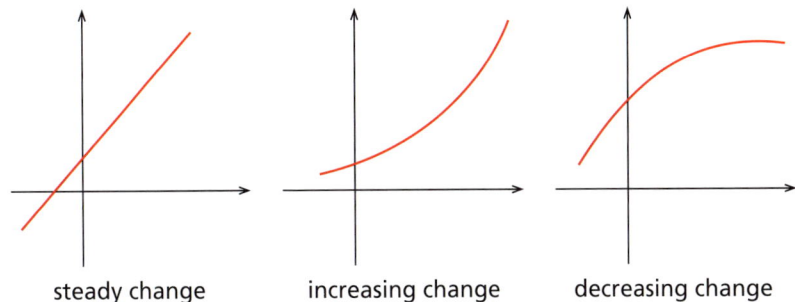

steady change   increasing change   decreasing change

Example   Water is poured at a steady rate into the conical glass shown. Sketch a possible graph of the depth of the water against time.

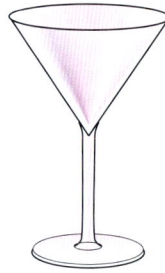

The bottom of the glass is narrow, so at the beginning the level is rising rapidly. As more water is added, the top of the water is wider, and hence the level rises less rapidly. The depth of water is increasing, but at a slower rate. The graph is curved downwards. The diagram shows a possible graph.

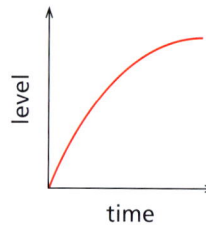

Exercise 10.5

1 The diagram shows four glasses, and four graphs of depth against time. Water is poured into each of the glasses at a steady rate. Which glass goes with which graph?

a   **b**   **c**   **d**

i)   ii)   iii)   iv)

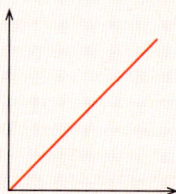   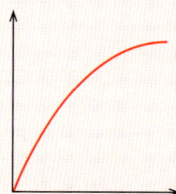   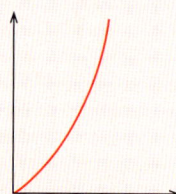   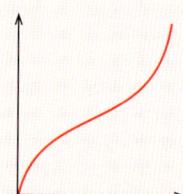

2 A politician claims that though prices are increasing, the rate of inflation is decreasing. Which of the graphs below would fit the claim?

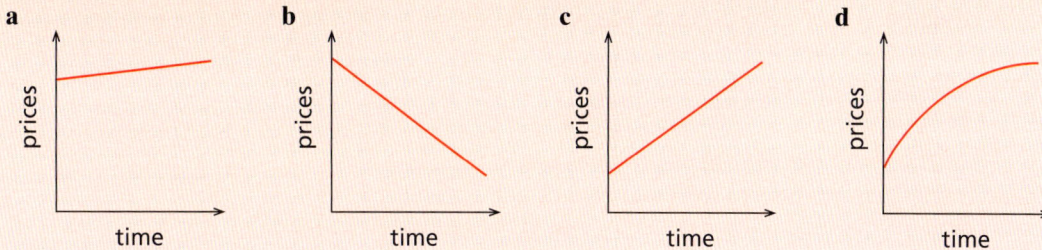

a

prices / time

b

prices / time

c

prices / time

d

prices / time

3 Cream is oozing steadily from the container shown. Sketch a graph for the level of cream against time.

4 Air is pumped at a steady rate into a spherical balloon. Sketch a graph for the radius of the balloon against time.

5 Water is pumped into a hemispherical bowl at a steady rate. Sketch graphs for
   a the depth of the water   **b** the width of the water surface.

6 A 'flu epidemic starts with only a few cases, spreads rapidly until a quarter of the population is affected, then gradually dies out. Sketch a graph showing the progress of the epidemic.

7 A third world country begins a campaign against illiteracy. Initially only 20% of the population are literate, but the government hopes that this will increase to 90% over 10 years. Sketch a graph showing the progress of the campaign.

8 An ecological organisation claims that destruction of the rain forest is continuing at an increasing rate. Sketch a graph to illustrate this claim.

9 An electrical circuit is switched on. Over 0.5 seconds the current increases until it reaches 10 amps. Sketch a graph for the current against time.

10 A children's book is written which becomes a classic. Sales rise to a peak and then fall to a steady level. Sketch a sales graph for the book.

SUMMARY

■ Functions which are **linear**, **quadratic** or **reciprocal** have distinctive **graphs**.
■ Quadratic graphs are in the shape of a **parabola**, with a **minimum** or **maximum** point.
■ Graphs are plotted by filling in a table of values. Be careful if the function is not defined at a particular value of x.
■ Equations can be solved by seeing where two graphs intersect.
■ If a quantity is increasing or decreasing at a rate proportional to itself, then it is **increasing** or **decreasing exponentially**. Its equation is of the form $y = ab^t$.

Exercise 10A

You will need:
• graph paper

1 Solve the equation $x^2 - 2x - 8 = 0$. Hence sketch the graph of $y = x^2 - 2x - 8$.

2 For what value of x is the function $y = \dfrac{2}{x+1}$ not defined? Sketch the graph of the function.

3 Plot the graph of $y = x^3 - x^2 + 1$, taking x values $-1, 0, 1, 2, 3$.

4 Use your graph for question 3 to solve the equation $x^3 - x^2 + 1 = 0$.

5 On your graph for question 3 draw the line $y = x$. Hence solve the equation

$$x^3 - x^2 - x + 1 = 0$$

6 Use a copy of a graph of $y = x^2$ to solve the equation $x^2 - x - 3 = 0$.

7 An electric current is switched off, and decreases exponentially. The graph shows the current over time. Find the equation of the curve, in the form $I = ab^t$.

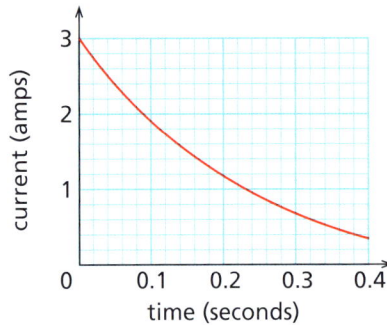

8 For the situation of question 7, find the level of the current after 0.5 seconds.

9 If a mortgage is paid off at a fixed rate, the greater part of the early repayments go towards interest on the debt. Over time, the proportion of the repayments which reduce the debt increases. Which of the graphs below describes the situation?

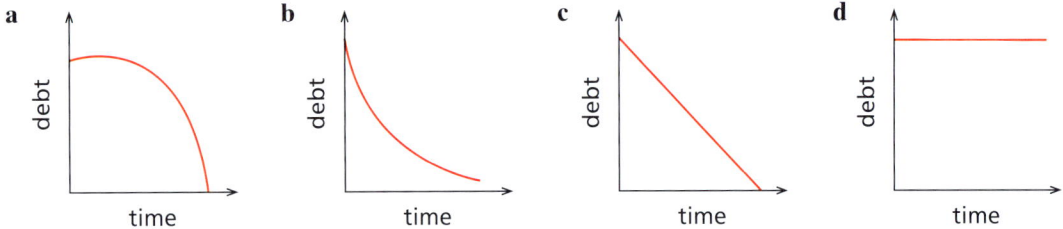

10 Water is poured at a steady rate into the 'waisted' glass shown. Sketch a graph of the water level over time.

Exercise 10B

1 Find the three points where $y = x^2 - 5x + 6$ crosses the axes. Hence sketch the curve.

2 Sketch the curve of $y = 1 + \dfrac{1}{x}$.

3 Plot the graph of the curve $y = x^2 - \dfrac{1}{x}$, for x between -2 and 2.

4 Use your graph for question 3 to solve the equation $x^2 - \dfrac{1}{x} = 1$.

5 On your graph for question 3 draw the line $y = x$. Hence solve the equation

$$x^3 - x^2 - 1 = 0$$

6 Make a copy of a graph of $y = x^2$ and use it to solve the equation $2x^2 + x - 1 = 0$.

7 The value of an investment is increasing exponentially, as shown in the graph. Find an equation of the curve in the form $V = ab^t$.

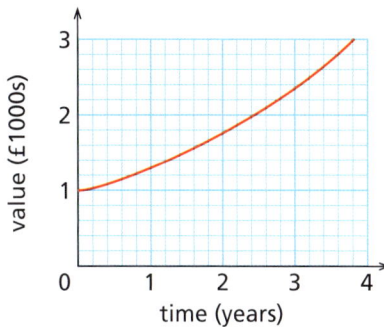

8 From your answer to question 7, find the value of the investment after 10 years.

9 The diagram shows a barrel of oil. A hole is made in the bottom, and the oil oozes out at a steady rate. Which of the graphs below shows the level of the oil against time?

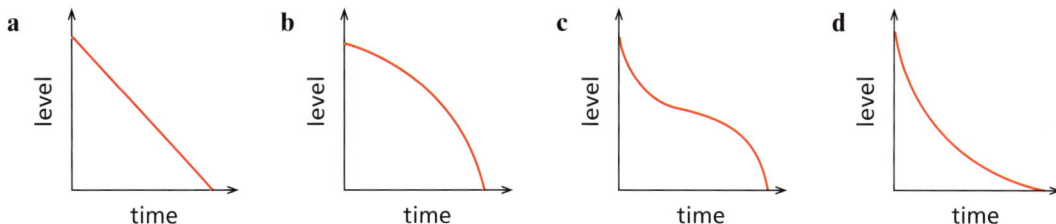

10 A politician claims that, although the crime rate is increasing now, the government's actions will bring about a decrease in a few years' time. On a copy of the diagram, sketch a graph to illustrate this claim.

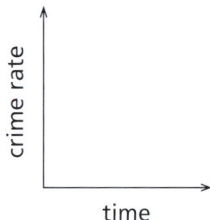

Exercise 10C Ma1

You will need:
* graph paper

Parabolas (quadratic graphs) have many practical uses. In particular, the reflecting surfaces of car headlights and radio telescopes have parabolic cross-sections. Here you find out why.

Plot as accurately as you can the graph of $y = x^2$. Label the point $(0, \frac{1}{4})$ as F.

Imagine a ray of light from F, striking the curve. Draw any line from F to the curve, meeting it at P. Draw, as accurately as you can, the tangent to the curve at P.

A ray of light from F will be reflected at P so that the angles with the tangent before and after reflection are equal, as shown. Draw the reflected ray of light.

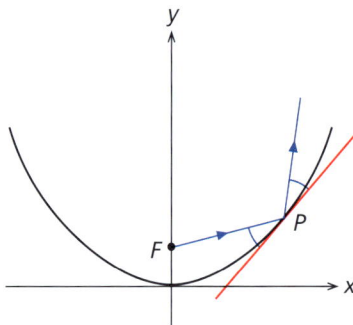

What do you notice? Test with some other lines from F.

What has this got to do with car headlights and radio telescopes?

Exercise 10D Ma1

You will need:
* graph paper

The economist Thomas Malthus, who lived from 1766 to 1834, is famous for his very pessimistic theory of development. He claimed that the population would grow exponentially, but that the food to support the population would only grow linearly. Hence the population would eventually outstrip the food supply and there would be starvation.

Suppose the population is P, and that the population who can be supported by the food available is Q. Malthus claimed that $P = ab^t$, but that $Q = mx + c$. Pick some suitable values for a, b, m and c, and plot the corresponding graphs. You should find that P will eventually overtake Q. It will help to use a graphics calculator for this.

In Britain, Malthus' gloomy predictions have not been fulfilled. Why not?

Exercise 10E (Ma1)

If a graph is curved, then its gradient is constantly changing. For a curved graph, the gradient at any point is the gradient of the tangent to the curve at that point. So to find the gradient of a curved graph, draw a tangent and find its gradient.

1 The diagram shows the graph of $y = x^2$.

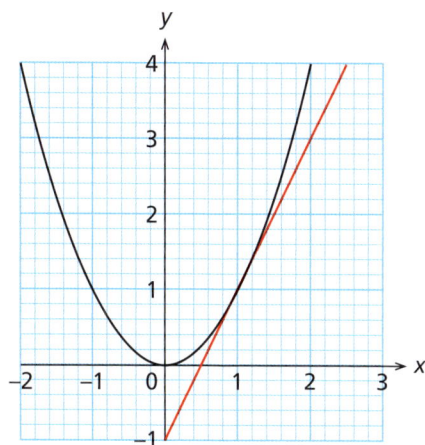

Make a copy of the graph.
  a Draw a tangent to the curve at $(1, 1)$. Find the gradient of this tangent.
  b Draw a tangent to the curve at $(-\frac{1}{2}, \frac{1}{4})$. Find the gradient of this tangent.
2 On page 142 there is the graph of $y = x^3$. Make a copy of the graph, and draw a tangent to find the gradient of the curve at $(1, 1)$.

Note. Finding gradients by tangents is very inaccurate. Other people may get different answers from yours. If you study maths at A level, you will meet an accurate method of finding the gradient of a curved graph, known as **differential calculus**.

11 : Travel graphs

Remember:

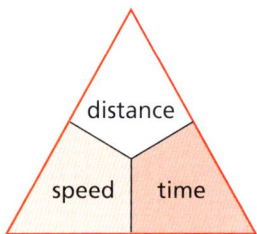

A graph can show how one quantity changes in terms of another quantity. In particular, it can show how distance or velocity change over time. It will then describe a journey. These graphs are **travel graphs**.

Distance time graphs

A **distance time graph** shows time along the horizontal axis and distance up the vertical axis. A distance time graph can describe a journey, in terms of how far was travelled and in what time.

The *speed* of a journey can also be found from the graph. It is given by the distance travelled divided by the time taken. This is the gradient of the graph.

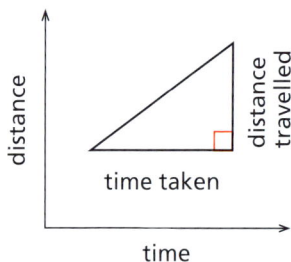

Examples

Mrs Cullinan goes to the market. The travel graph below shows her distance away from home. Describe her journey. What was her speed in the first part of the journey? Give your answer in miles per hour.

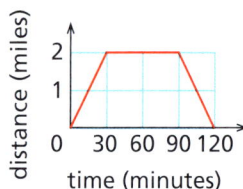

The graph goes through $(0, 0)$, $(30, 2)$, $(90, 2)$ and $(120, 0)$. So Mrs Cullinan walks two miles in 30 minutes, then spends 1 hour at the market, then walks home in 30 minutes.

During the first part of the journey she walked 2 miles in 30 minutes, that is in $\frac{1}{2}$ hour. Her speed was 4 m.p.h.

Kevin is going to a football match 10 miles away. He cycles for 30 minutes at 12 m.p.h., then stops for 20 minutes to chat with a friend, then cycles the remaining 4 miles at 10 m.p.h. Draw a travel graph to show his journey.

For the first part, he travels $12 \times \frac{1}{2}$ miles, that is, 6 miles. Join up $(0, 0)$ and $(30, 6)$. The graph is then flat, so join up to $(50, 6)$. For the final stretch he cycles at 10 m.p.h. for 4 miles, so the time taken is $\frac{4}{10}$ hour. Convert to minutes by multiplying by 60.

$$\frac{4}{10} \times 60 = 24$$

So the final stretch took 24 minutes. Join up to (74, 10). The completed graph is shown.

Exercise 11.1

1   Mr McCluskey is an estate agent. On a certain morning he has to visit two clients. This travel graph shows his journey. Describe the journey, giving his speed on the first stage.

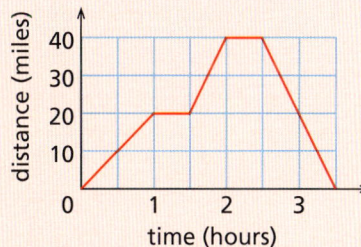

2   Theresa is a farmer. She drives to the nearest town to buy seed. Describe her journey, giving the speed on the final stage.

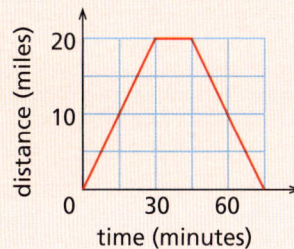

3   Mr Seger leaves work, goes to a club to play squash, then returns home. Describe the journey, giving the speeds during each stage.

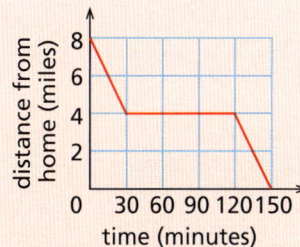

4   Andrea walks to a shop, covering 2 miles in 30 minutes. She spends 10 minutes in the shop, then walks back in 40 minutes. Draw a travel graph to show the journey.

5   Mr Rahman walks to the bus stop, taking 15 minutes to cover 1 mile. He waits 10 minutes, then gets on the bus which takes 30 minutes to cover 6 miles. He then walks the remaining $\frac{1}{2}$ mile to work in 10 minutes. Draw a travel graph to show the journey.

6   Mrs Sinclair goes swimming before work. She walks two miles to the pool, at 4 miles per hour, then spends 40 minutes at the pool, then walks back one mile in 20 minutes to her place of work. Draw a travel graph to show her journey.

Intersecting travel graphs

Suppose you draw travel graphs for two people travelling along the same road. When the two graphs cross, that means that the people have met each other.

Example Asil drives North from London along the M1, starting at noon. He drives at a steady speed of 60 m.p.h. Half an hour later Bernard sets off in pursuit, travelling along the same road at a steady speed of 70 m.p.h. When does Bernard catch up with Asil? How far from London are they when this happens?

Draw lines on a travel graph to show their journeys. Asil starts at the origin, and after 1 hour he has driven 60 miles. So plot (1, 60) on the graph. Join this up to the origin by a straight line. Bernard starts later, so his line begins at 0.5 on the horizontal axis. After 1 hour he has travelled 70 miles, so plot (1.5, 70) on the graph. Join these points up by a straight line. The lines cross at (3.5, 210). So Bernard catches up with Asil after 3.5 hours, and they are 210 miles North of London at that time.

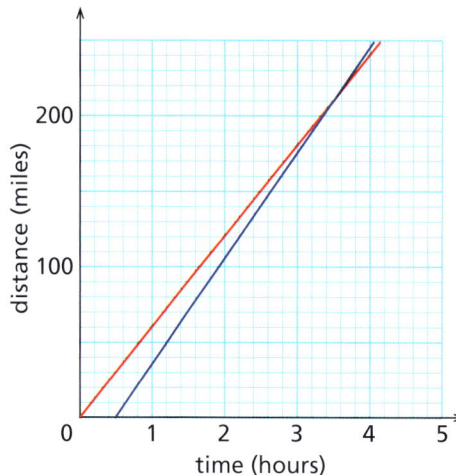

Exercise 11.2

1 John sets off for school at 8 o'clock, walking at 3 m.p.h. Twenty minutes later his sister notices that he has left his books behind, so she runs after him at 6 m.p.h. Copy the diagram and draw travel graphs to show their journeys. When does his sister catch up with him? How far are they from home when this happens?

2 Mike's school is 800 m from his home. At 4 o'clock he sets off from school to home, walking at a speed of 50 m per minute. At the same time, his sister Deirdre leaves home, walking towards the school at 100 m per minute. Copy the diagram and draw travel graphs to show their journeys. When do they meet? How far are they from home when they meet?

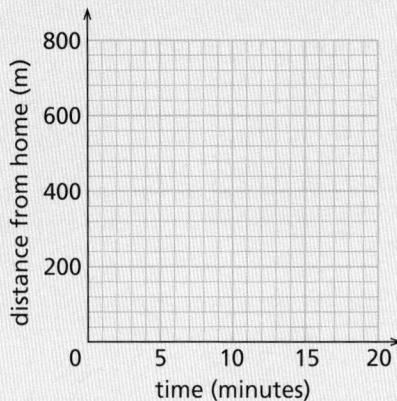

3 A balloon is 1000 m up in the air, and starts descending at 10 m/s. At the same time another balloon starts rising from the ground at 15 m/s. Copy the diagram and draw travel graphs to show their heights above the ground. When are they at the same level?

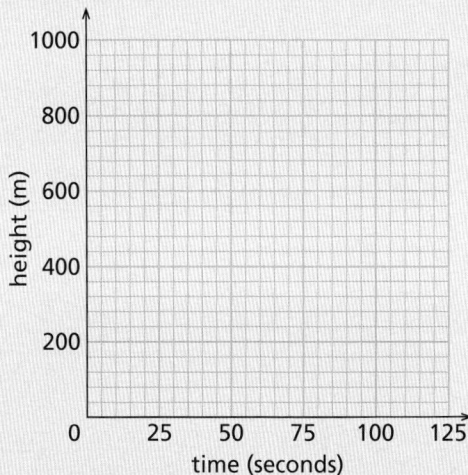

4 A coach leaves Bristol, travelling East along the M4 at 50 m.p.h. Half an hour later a car leaves Bristol, travelling East at 70 m.p.h. When does the car overtake the coach, and how far are they from Bristol at the time?

5 A plane leaves an airport, flying North at 400 m.p.h. Half an hour later another plane leaves the airport, flying North at 500 m.p.h. When does the second plane overtake the first, and how far North have they flown?

6 London is 3500 miles due North of Accra (the capital of Ghana). A plane leaves London, flying South at 500 m.p.h. At the same time a plane leaves Accra, flying North at 600 m.p.h. When do the planes pass each other, and how far are they from London at this time?

Curved graphs

The travel graphs so far have all consisted of straight line segments. That means that in each section of the journey the speed was constant. This rarely happens in any real-life journey. The speed of a walker, a cyclist or a car is constantly changing. The travel graph for a journey with changing speed is curved.

If an object's speed is increasing, the object is **accelerating**. If the object's speed is decreasing, the object is **decelerating**.

Graphs for acceleration and deceleration are shown below. For the acceleration graph on the left, notice that the graph is getting steeper, so the slope is increasing. For the deceleration graph on the right, notice that the graph is getting shallower, so the gradient is decreasing.

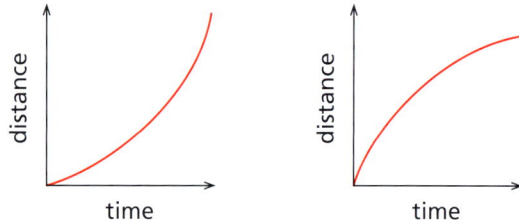

Example   A car accelerates from rest to a constant speed, then holds that speed for a short while, then brakes to a halt. Sketch a possible travel graph for the journey.

The speed increases initially, so the graph curves upwards. During the steady speed period the graph is a straight line. During the braking period the graph curves downwards, eventually becoming flat.

The graph is shown in the diagram.

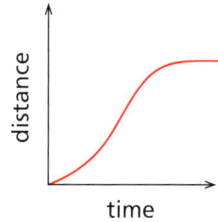

Exercise 11.3

1 The graph shows the journey of a bus between two stops. Describe the journey.

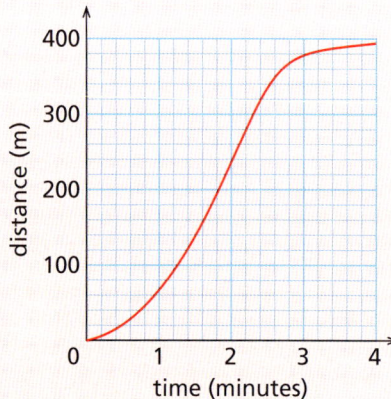

2 The graph shows the distance travelled by a sprinter in a race. Describe his progress.

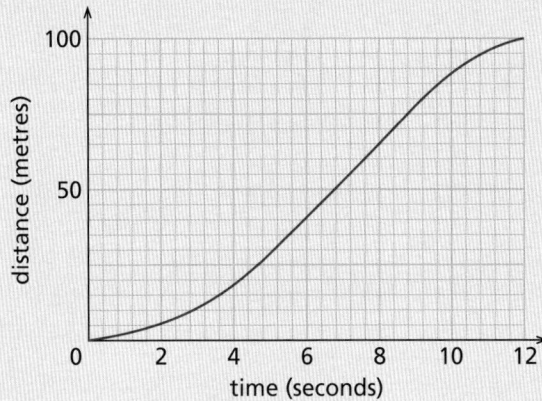

3 Two runners, Sandra and Betty, entered a 100 m race. The diagram shows their progress.
  a Who won the race?
  b Describe the race.

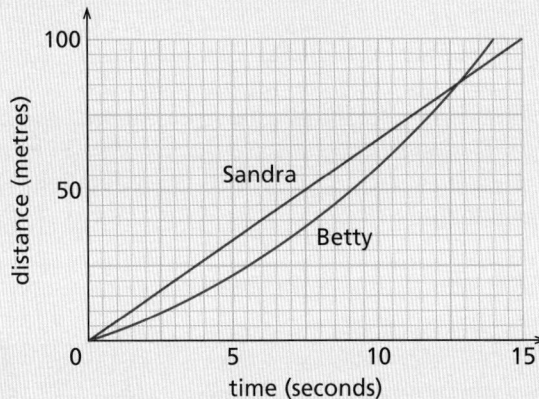

4 A parachutist jumps out of a plane, falls freely for a few seconds, then the parachute opens and he descends to the ground. Sketch a possible travel graph for the journey.

5 A swimmer dives from the high board, descends under water and then resurfaces. Sketch a possible travel graph for her dive.

6 In a horse race, Gingerbeer ran at a constant speed. Kismet started more slowly, kept behind for most of the race but sprinted at the end to win the race. Sketch a possible travel graph to show the progress of the race.

7 In a 400 m race, Ben started rapidly but then slowed down towards the end. Michael ran at a constant speed, and the two runners finished the race simultaneously. Sketch a possible travel graph to show the progress of the race.

8 A bank robber runs away from the scene of the crime at a constant rate. A police car accelerates until it catches up with the robber. Sketch a possible travel graph to show the chase.

Increasing velocity is acceleration.
Decreasing velocity is **deceleration**.
Deceleration can be thought of as negative acceleration.
deceleration of $2\,m/s^2$ = acceleration of $-2\,m/s^2$

Velocity time graphs

In our travel graphs so far, we have shown distance along the vertical axis. A graph with **velocity** on the vertical axis is a **velocity time** graph.

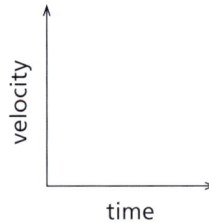

Here are some simple velocity time graphs, for a car journey.

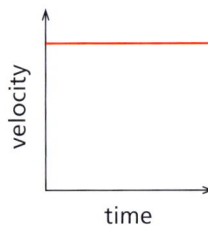

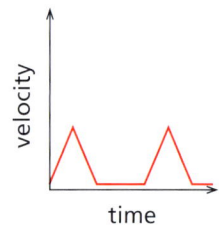

In the first graph, the velocity is constant. So the car (or person, or train) is travelling with a fixed velocity. The distance is changing.

In the second graph, the velocity is decreasing to zero. So the car is slowing to a halt. But the car is moving forwards throughout the motion.

In the third graph the velocity increases, then decreases down to zero. The velocity is zero for a while, then increases again and decreases back down to zero. This might represent a car stuck in a traffic jam.

The rate of change of velocity is **acceleration**. So the gradient of a velocity time graph, the rate of change of the velocity, gives the acceleration.

Examples The diagram below shows the velocity time graph for the journey of a car. Describe the journey.

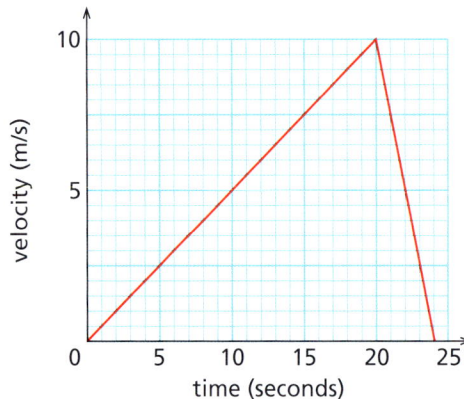

The car starts from zero velocity, then accelerates uniformly to a maximum velocity of 10 m/s, then brakes uniformly to a halt.

Find the acceleration for each part of the journey on page 162.

The acceleration for the first part of the journey is the slope of the first line. Divide the change of speed by the time taken.

$$\text{acceleration} = \frac{\text{final speed} - \text{initial speed}}{\text{time taken}} = \frac{10}{20} = 0.5$$

Similarly, for the second half, the acceleration is $-\frac{10}{4}$.
    The acceleration in the first part is $0.5\,\text{m/s}^2$.
    The acceleration in the second part is $-2.5\,\text{m/s}^2$.

> For the second part, we could say that the car was decelerating at $+2.5\,\text{m/s}^2$.

A train travels at a steady 30 m/s for 20 seconds, then brakes to a halt over a period of 10 seconds. Draw a velocity time graph for this journey.

Draw a horizontal line at $v = 30$, for $t = 0$ to $t = 20$. After a total time of 30 seconds the train has halted, so join the endpoint to $(30, 0)$. The graph is as shown.

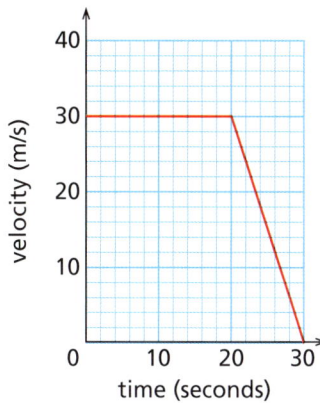

Exercise 11.4

1 The diagram shows the velocity of a bus over a period of 40 seconds.
  a Describe the journey.
  b What was the acceleration of the bus during the first part of the journey?
  c What was the deceleration of the bus during the last part of the journey?

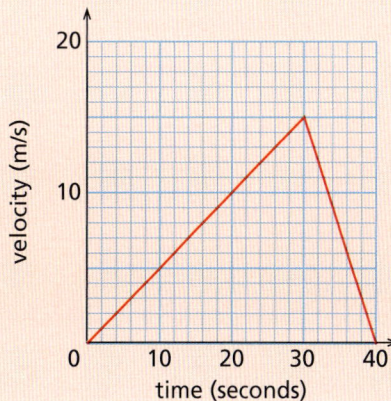

2 The diagram shows the velocity of a car as it entered a village with a speed restriction.
  a Describe the journey.
  b What was the deceleration of the car during the first part of the journey?
  c What was the acceleration of the car during the last part of the journey?

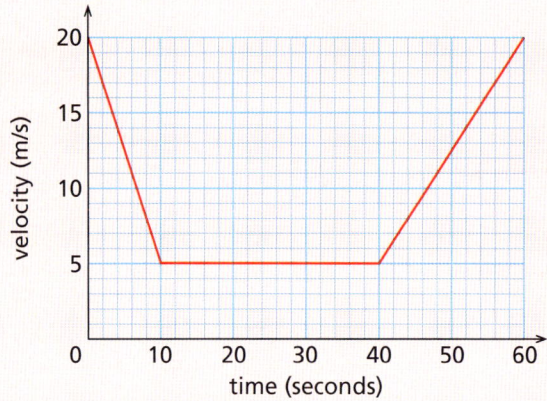

3 The journey of a multi-stage rocket is shown in the diagram.
  a Describe its ascent.
  b Find its acceleration for each of the three stages of the ascent.

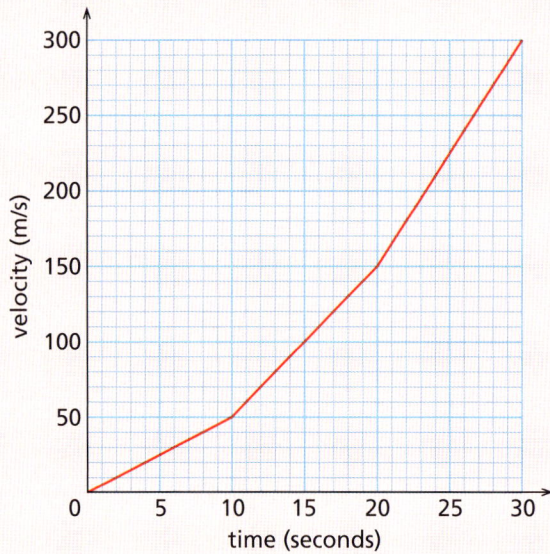

4 The velocity of a tube train between stations is shown. Find the acceleration for the first part of the journey and the deceleration for the last part.

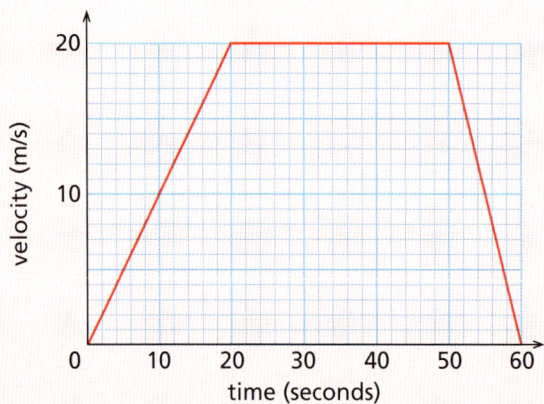

5 A car accelerates at 2 m/s for 10 seconds, holds a constant speed for 20 seconds, then brakes uniformly to a halt over 5 seconds. Draw a velocity time graph for its journey.

6 A cyclist accelerates uniformly to 10 m/s over 5 seconds, then holds that speed for 20 seconds, then brakes uniformly to a halt over 2 seconds. Draw a velocity time graph for the journey.

Area under velocity time graphs

Look at this velocity time graph. It shows a journey, at a constant 12 m/s over 10 seconds. Notice two things.

The distance travelled is 12×10 m, which is 120 m.
The region under the graph is a rectangle, of area 12×10, that is, 120.

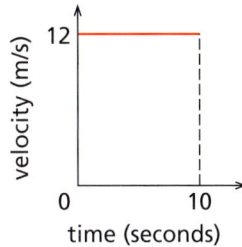

These are the same. This is always true, even if the graph is not a horizontal line.

● *The area under a velocity time graph gives the distance travelled.*

Justification
The diagram shows a velocity time graph. The velocity is changing, but in each little strip the velocity is almost constant. The area in each strip is equal to the horizontal distance, representing time, multiplied by the vertical height, representing velocity.

area of strip = time $\times$ velocity = distance

The area in each strip represents the distance travelled during a small time period, so the total area represents the total distance travelled.

Example   Look at the example on page 162, about the velocity of a car. Find the distance travelled by the car.

The distance is given by the area of the graph. The region under the graph is a triangle, with base 24 seconds and height 10 m/s. This has area

$$\tfrac{1}{2} \times 24 \times 10 = 120$$

The distance travelled is 120 m.

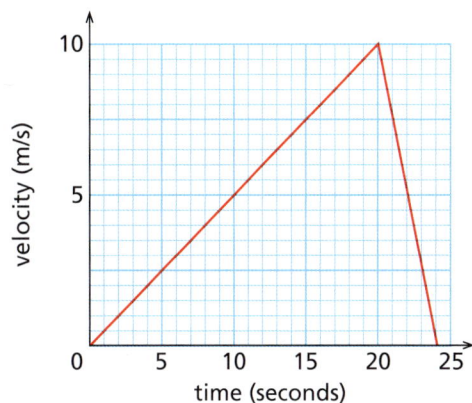

Exercise 11.5

1 Look at the diagram below. Find the distance travelled by the bus.

2 Look at the diagram below. Find the distance travelled by the car.

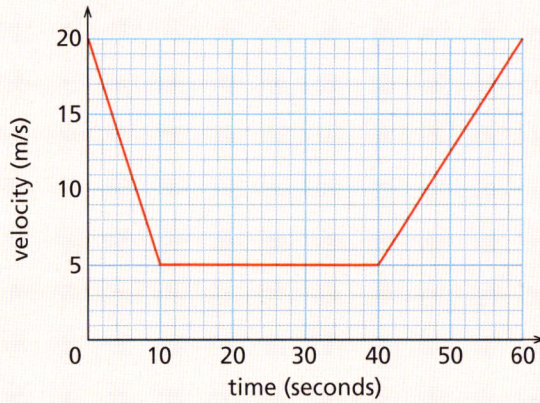

3 Look at the diagram below. Find the height reached by the rocket.

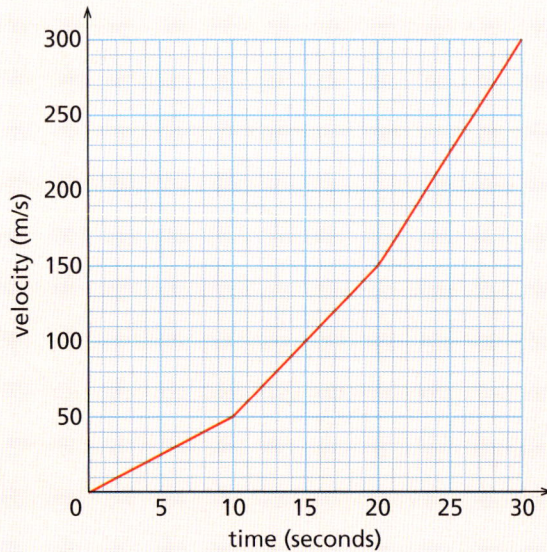

4 Look at the diagram below. Find the distance between the stations.

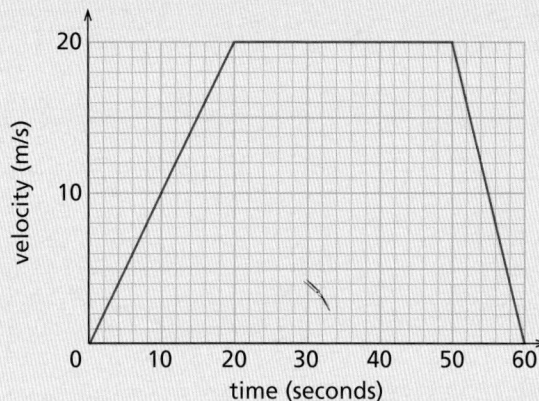

5 During take-off, a plane accelerates uniformly to a speed of 100 m/s over a time of 30 seconds. Sketch a velocity time graph, and find the distance travelled during take-off.

6 A plane comes in to land on an aircraft carrier. It arrives at 50 m/s, and comes to a halt within 10 seconds. Sketch a velocity time graph, assuming uniform deceleration. What is the least possible length of the flight deck?

7 A car accelerates uniformly to a speed of 20 m/s, over a period of 10 seconds. It holds that speed for 30 seconds, then decelerates uniformly to a halt over 10 seconds. Sketch a velocity time graph for the journey, and find the distance travelled.

8 On entering a village, a car decelerates from 20 m/s to 10 m/s, over a period of 5 seconds. It maintains a constant speed for 10 seconds, then accelerates up to 20 m/s over a period of 10 seconds. Sketch a velocity time graph, and find the total distance travelled during this journey.

9 A train accelerates at 0.2 m/s² over a period of 100 seconds, then holds a constant velocity for 200 seconds, then brakes uniformly to a halt over 50 seconds. Sketch a velocity time graph and find the total distance travelled.

10 A car is driven at a constant 20 m/s for 30 seconds, then it accelerates uniformly to 40 m/s over a period of 20 seconds. Sketch a velocity time graph and find the total distance travelled.

11 A car accelerates uniformly from rest at 1.5 m/s², covering a distance of 48 m. Find the time taken.

12 A train accelerates uniformly to 40 m/s over 20 seconds, travels at a constant speed, then brakes to a halt over 10 seconds. The total distance travelled is 1600 m. What is the total time of the journey?

13 A car accelerates to 20 m/s, holds that speed for 40 seconds, and then decelerates to a halt. The total distance covered is 1000 m. The acceleration and deceleration are uniform and of equal rate. Find the total time taken.

SUMMARY

- A **distance time graph** has time along the horizontal axis and distance up the vertical axis. The **gradient** of a section of a distance time graph gives the speed.
- The **intersection** of two distance time graphs can show when two people or vehicles meet.
- A journey in which the speed is changing has a curved distance time graph.
- A **velocity time graph** has time along the horizontal axis and **velocity** up the vertical axis. The gradient of a velocity time graph gives the **acceleration**. The area under a velocity time graph gives the distance travelled.

Exercise 11A

1 The Milne family drives along the motorway, stopping at a service station during the journey. The diagram shows a distance time graph of the journey.
   a How long was spent at the service station?
   b What was the speed of the car in the final part of the journey?

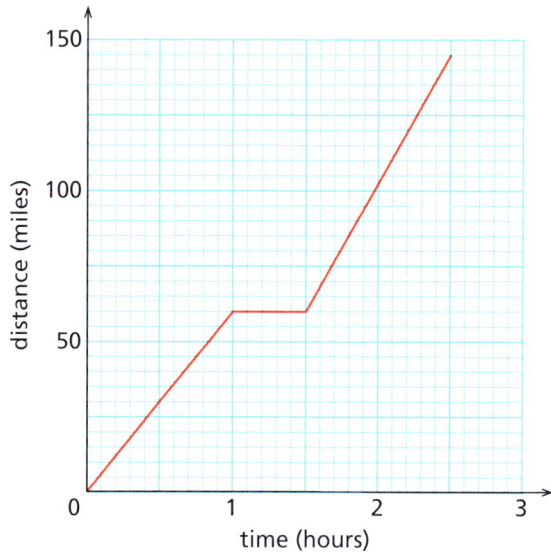

2 Gordon walks at 4 m.p.h. for 30 minutes, then rests for 10 minutes, then walks back at 3 m.p.h. Draw a distance time graph for the journey.

3 The diagram shows the height of a rocket above the ground. Describe the journey of the rocket.

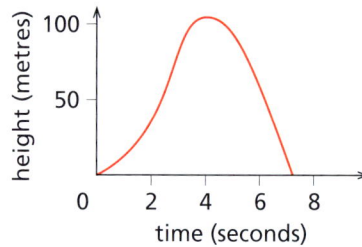

4 A resting cheetah sees a gazelle running away from it. The cheetah sets off in pursuit, but after 10 seconds realises it will not be able to catch up with the gazelle and gives up. Sketch a possible distance time graph for the cheetah.

5 The diagram shows the velocity of a car.
   a What is the acceleration for the first part of the journey?
   b What is the deceleration for the final part of the journey?
   c What is the total distance travelled?

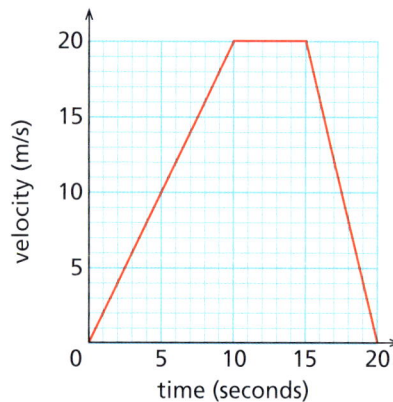

6 A ship leaves port, accelerating uniformly to 10 m/s over a period of 30 seconds. It holds that speed for 60 seconds. Draw a velocity time graph for this section of the ship's voyage.

7 A car accelerates at 2 m/s² for 20 seconds, holds a constant speed for 30 seconds, then brakes uniformly to a halt. The total distance travelled was 1760 m. Find the deceleration during the braking period.

Exercise 11B

1 A dog owner throws a stick for his dog to fetch. The diagram shows the journey of the dog.
  a What was the speed of the dog when running towards the stick?
  b What was the speed of the dog when running back?

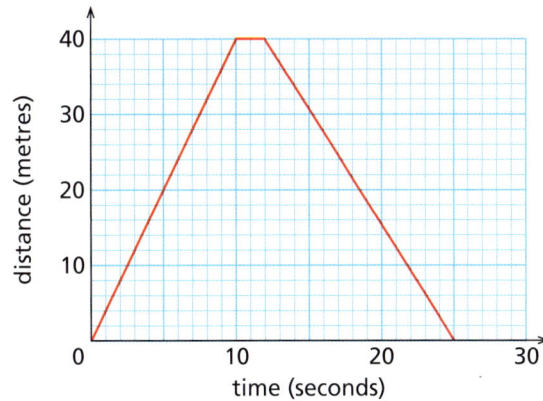

2 The diagram shows the distance of a rocket from the Earth as it travels to the Moon. Describe the journey.

3 The diagram shows part of a roller coaster ride. The cars start from rest at *A*. Sketch a possible distance time graph for the *horizontal* distance travelled by the cars. Indicate *A*, *B* and *C* on your graph.

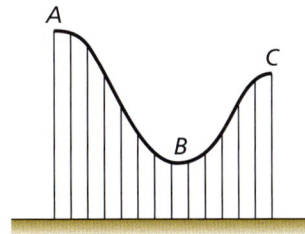

4 A swimmer dives from the high board into a swimming pool. The diagram shows her velocity.
  a What was her acceleration downwards while she was in the air?
  b What was her deceleration downwards while she was in the water?
  c How high was the board above the water?
  d How deeply did she dive into the water?

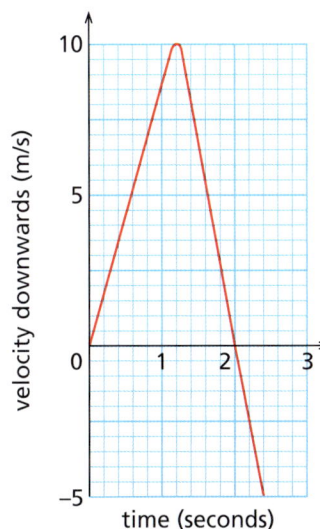

5 A ball is thrown upwards at 40 m/s. It decelerates uniformly to a halt after 4 seconds, then falls to the ground, reaching it after a further 4 seconds. Draw a velocity time graph for the ball's travel.
6 A car accelerates uniformly at 2 m/s², covering a distance of 36 m. For how long was it accelerating?

Exercise 11C

The diagram shows a distance time graph for a horse race with four horses. Provide a racing commentary.

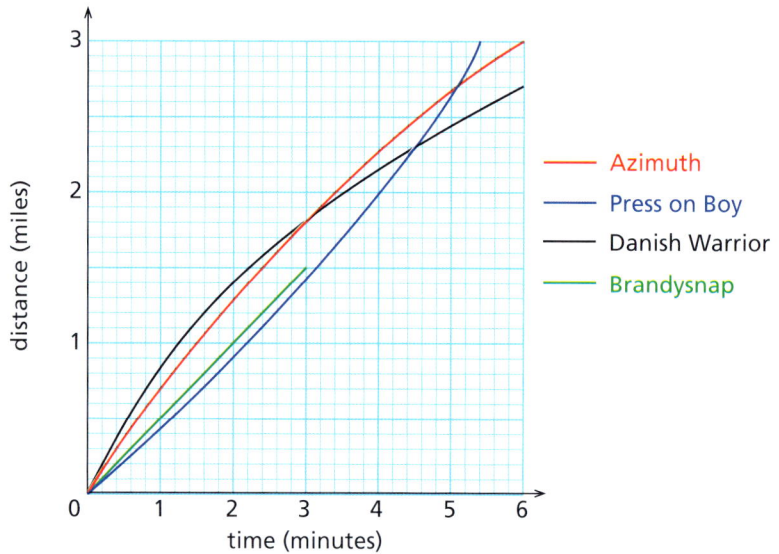

- Azimuth
- Press on Boy
- Danish Warrior
- Brandysnap

Exercise 11D

1 The diagram shows a distance time graph for a journey of a car. Sketch a velocity time graph for the same journey.

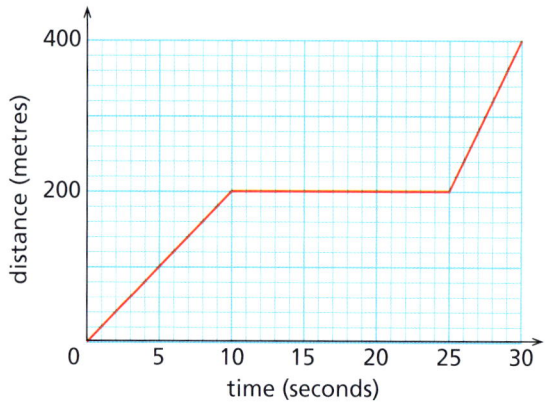

2 The diagram shows a velocity time graph for a journey of a train. Sketch a distance time graph for the same journey, given that it started at zero distance.

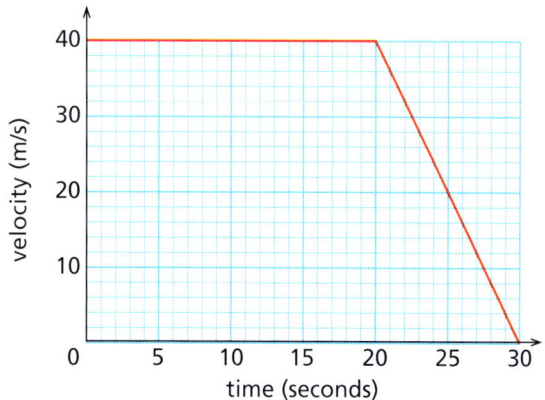

Exercise 11E

In the final section of this chapter you found the area under a velocity time graph, which gives the distance travelled. If the graph consists of straight line segments, this can be done by finding the areas of triangles and trapezia. What if the graph is not a straight line, that is, if the acceleration is changing? The following is a way to approximate the area under a curve. It is called the **trapezium rule**.

Divide the x-axis into equal lengths of h. Let the y-values be $a_1, a_2, a_3, \ldots, a_n$.

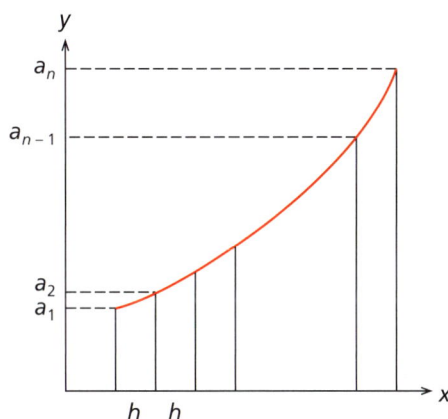

Then the area under the curve is approximately

$$h(\tfrac{1}{2}a_1 + \tfrac{1}{2}a_n + a_2 + a_3 + \ldots + a_{n-1})$$

> So you add the end values halved, then all the middle values, then multiply by h.

The larger n is, the better the approximation.

This involves a lot of calculation. A spreadsheet can do all the number crunching. Suppose we want to find the area under the curve $y = x(4 - x)$, between $x = 0$ and $x = 4$.

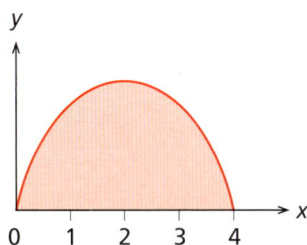

Divide the x interval into 40 equal segments of length 0.1 each, so $h = 0.1$. In the A column of a spreadsheet put the endpoints of these intervals, so enter 0 in A1 and =A1+0.1 in A2. Copy down to A41.

> You could use the 'fill' command.

In the B column, put the values of y. So put =A1*(4−A1) in B1, and copy down to B41. You can use the =SUM function to add the middle values. Enter =SUM(B2..B40) in C1. Add this to half B1 and half B41. Multiply by 0.1, and this gives the approximate area.

Adapt your spreadsheet for a smaller value of h, that is, a larger number of segments. What is the approximate area tending to?

12 Accuracy

We cannot measure things like lengths, weights, times and so on with perfect accuracy. Our results are always slightly different from the true value. The difference is the **error**. So the important thing is to be aware of how inaccurate an answer is, and not to give it to too high a degree of accuracy.

Error

The error of a measurement is the difference between the true value and the measured value. Suppose the true speed of a car is 43 m.p.h., but the speedometer measures it as 39 m.p.h. The error is 4 m.p.h. Similarly, if the speedometer measures the speed as 45 m.p.h., then the error is 2 m.p.h.

Note that the error is always positive. We take the *positive* difference between the two values, whether the measurement is greater or less than the true value.

> This is sometimes called the **absolute** error.

The importance of an error depends on how large the true value is. Obviously, an error of 0.1 second is more important when timing a 100 m race than when timing a marathon! The relative error compares the error with the true value.

$$\text{relative error} = \frac{\text{error}}{\text{true value}}$$

We might not know the true value. In this case, divide the error by the measured value.

$$\text{relative error} = \frac{\text{error}}{\text{measured value}}$$

> This is the same as converting fractions to percentages.

Sometimes the relative error is expressed as a percentage. It expresses the error as a percentage of the true value.

$$\text{percentage error} = \text{relative error} \times 100 = \frac{\text{error}}{\text{true value}} \times 100$$

Example Jamie knows that his weighing machine can be inaccurate by up to 2 kg. He weighs himself as 80 kg. What is the maximum percentage error in the measurement?

We do not know the true weight. To find the relative error, divide the error by the measured weight.

$$\text{relative error} = \frac{2}{80} = 0.025$$

Convert to a percentage.

$$\text{percentage error} = 100 \times \text{relative error} = 100 \times 0.025$$

The maximum percentage error is 2.5%.

Exercise 12.1

1 John's true weight is 73 kg. His scales show 71 kg. What is the error in the weight?
2 The distance between two towns is 56 miles, but a car's odometer measures it as 58 miles. What is the error in the measurement?
3 A time is measured as 35 seconds, but the error could be as much as 2 seconds either way. Between what limits does the true time lie?
4 The temperature of a bathing pool is measured at 66 °F, but it could be out by up to 3 °F. Between what limits does the true temperature lie?
5 To convert temperatures from Celsius to Fahrenheit, the exact formula is $F = \frac{9}{5}C + 32$. An approximate formula, which is easier to use, is $F = 2C + 30$. What is the error in using this formula when
   a $C = 10$ b $C = 30$ c $C = 50$?
   For what value of C does the approximate formula give the correct result?
6 An approximate conversion between Imperial and metric units is

    Two and a quarter pounds of jam weighs about one kilogram   1 lb = 0.452 kg

   What is the error in converting 10 lb to kilograms using this rule?
7 Patricia is on holiday in Australia, and wants to know how many Australian dollars she will get for £480. She uses the approximation that £1 is worth about 2.5 dollars. Find the error in using this rule when the exact rate of exchange per £ is
   a $2.5632   b $2.4355.
8 An approximate rule for changing miles to kilometres is that 1 mile = 1.5 km. A more exact rule is that 1 mile = 1.6093 km. What is the error in using the approximate rule when
   a changing 80 miles to kilometres   b changing 60 km to miles?
9 An error of 1 kg is made when weighing an object. What is the relative error in these cases?
   a The object is a parcel weighing 8 kg.
   b The object is a lorry weighing 3200 kg.
10 An error of 1 m is made when measuring a distance. What is the relative error in these cases?
   a The distance is the 5 m width of a room.
   b The distance is the 120 m length of a playing field.
11 An error of 3 volts is made when measuring a voltage. Find the percentage error in these cases.
   a The voltage is the 12 volts of a car battery.
   b The voltage is the 240 volts of mains electricity supply.
12 A weight of 8 kg is measured as 7.5 kg. Find
   a the error   b the relative error   c the percentage error.

13 A time of 120 seconds is measured as 123 seconds. Find
  a the error
  b the relative error
  c the percentage error.

14 An electric current of 5.34 amps is measured as 5 amps. Find
  a the error
  b the relative error
  c the percentage error.

15 A volume of 1.156 litres is measured as 1.1 litres. Find
  a the error
  b the relative error
  c the percentage error.

16 A weighing machine for postage is inaccurate by up to 2 grams. Find the maximum relative error if
  a the machine weighs a letter as 20 grams
  b the machine weighs a parcel as 250 grams.

17 Jonathan reckons he can time a race to within 0.1 seconds. What is the maximum relative error if
  a he times a 100 m sprint as 12.4 seconds?
  b he times a 1000 m race as 4 minutes 38.2 seconds?

18 The maximum *relative* error of a weighing machine is 0.02. Find the maximum error if the machine weighs
  a a parcel as 800 grams   **b** a letter as 28 grams.

19 When Trudie measures a distance, she knows that her percentage error can be up to 3%. Find the maximum error when she measures
  a the height of a room as 2.8 m
  b the length of a garden as 12.4 m.

20 The maximum percentage error of a stopwatch is 0.15%. Find the maximum error if it measures
  a the time of a 100 m race as 14.6 seconds
  b the time of a 1000 m race as 6 minutes 12.2 seconds.

21 The maximum percentage error of a motorbike's speedometer is 4%. Find the maximum error if it registers a speed of 60 m.p.h.

22 A decorator converts 16 feet to metres, and makes an error of 0.1088 m. What approximation has he been using?

> 1 foot = 0.3068 m

23 Kimiko converts ¥100 000 to £, and makes an error of £46. If the true rate is ¥183 per £, what approximation has she been using?

24 A temperature of 20 °C is measured as 21 °C. Does it make sense to find the relative error?

Rounding errors

When we round a measurement there is automatically an error. The size of the error depends on how we have rounded.

Suppose we round to the nearest whole number. Then the original number could be up to 0.5 on either side of the whole number.

Suppose we round to two decimal places. Then the original number could be up to 0.005 on either side of the rounded number.

Suppose we round to the nearest 5. Then the original number could be up to 2.5 on either side of the rounded number.

In general, suppose we round correct to a certain unit. Then the original number could be up to half that unit on either side of the rounded number.

Note. The values that the original number could take are not symmetrical about the rounded number. Suppose that a number is given as 7, correct to the nearest whole number. Then the original number must be *less* than 7.5, but *greater than or equal to* 6.5. If x is the original number, then

$$6.5 \le x < 7.5$$

In terms of a number line, the region containing x is shown in the diagram. Note that there is a solid dot at 6.5 (containing 6.5 itself) but a hollow dot at 7.5 (not containing 7.5 itself).

Exercise 12.2

1 A length is given as 5.3 cm, correct to one decimal place. Find the values between which the true length lies.
2 A time is given as 12.1 seconds, correct to one decimal place. Find
 a the values between which the true time lies
 b the maximum error of the measurement
 c the maximum relative error of the measurement.
3 A weight is given as 2.34 kg, correct to two decimal places. Find
 a the values between which the true weight lies
 b the maximum error of the measurement
 c the maximum relative error of the measurement.
4 Suppose y has been rounded to 21, correct to the nearest whole number. Write down, using the $\le$ and $<$ symbols, the limits between which y lies.
5 On a copy of this number line, indicate the limits of the number y of question 4.

6 A number z has been rounded to 3.23, correct to two decimal places. Write down, using the $\le$ and $<$ symbols, the limits between which z lies.
7 On a copy of this number line, indicate the limits of the number z of question 6.

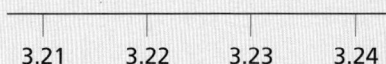

8 For each of the following, give the limits between which the number lies. In each case indicate the limits on a number line.
 a p, rounded to 270, to the nearest 10
 b q, rounded to 17, to the nearest whole number
 c r, rounded to 2.1, correct to one decimal place
 d s, rounded to 0.18, correct to two decimal places
 e t, rounded to -1.244, correct to four significant figures
 f u, rounded to 0.000 157, correct to three significant figures
 g v, rounded to -0.0026, correct to two significant figures
9 A weight is given as 155 lb, correct to the nearest pound, and as 70 kg, correct to the nearest kilogram. Between what limits does the true weight lie? Give your answer in kilograms.

1 lb = 0.4536 kg

Combining measurements

When we combine measurements, the errors are also combined.

Adding and subtracting

When two quantities are added, the maximum errors are added. Suppose we are adding two lengths, by putting two sticks together end to end. The first stick should be 2 m long, and the second should be 1 m long. If the first stick is x m too long, and the second stick is y m too long, then the combined length is $(x + y)$ m too long.

$$(2 + x) + (1 + y) = 3 + (x + y)$$

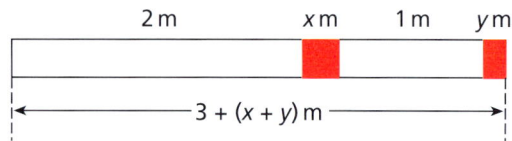

When two quantities are subtracted, the errors are added. Suppose we are finding the difference between two lengths, by putting one stick alongside another. Suppose the first stick is x m too long, and the second stick is y m too *short*. Then the difference between the sticks is $(x + y)$ m too great.

$$(2 + x) - (1 - y) = 1 + (x + y)$$

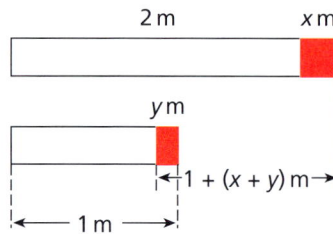

So when adding or subtracting measurements, the errors are added.

Example

A teacup holds 110 cm³ of tea, correct to the nearest 10 cm³. When taking the cup to the table, 20 cm³, correct to the nearest 5 cm³, is spilled into the saucer. What is the least amount of tea left in the cup?

In the original amount of tea, the maximum error is 5 cm³. In the amount that spilled out, the maximum error is 2.5 cm³. Add these errors, and subtract from 90 cm³.

> **Remember:**
>
> *If a quantity is given correct to a certain unit, the error is up to half that unit.*

maximum error = 7.5 cm³
least amount remaining = (110 − 20 − 7.5) cm³

The least amount remaining is 82.5 cm³.

Note. This example shows why the errors must be added, even when the quantities are subtracted. The least amount of tea is left when the cup holds the *least* amount originally, and the *greatest* possible amount is spilled.

Exercise 12.3

1 Jane is 160 cm tall, and John is 165 cm tall, both being given to the nearest centimetre. What is the maximum error in
  a their total height?
  b the difference in their heights?

2 Two runners in a sprint relay race took 20.1 seconds and 22.5 seconds respectively, both being given to one decimal place. What is the maximum error in
  a the total time they took?
  b the difference in their times?

3 The weight of a lorry is 4200 kg, and the weight of its load is 3500 kg, both figures being given to the nearest 100 kg. What are the limits between which the total weight lies?

4 The weight of a horse is 840 kg, correct to the nearest 10 kg, and the weight of its rider is 75 kg, correct to the nearest 5 kg. What are the limits between which the total weight lies?

5 The area of a garden is 120 m², correct to the nearest 10 m², and the ground area of the house is 65 m², correct to the nearest 5 m². What are the limits between which the total area of the property lies?

6 The total time of a television news programme was 18 minutes, correct to the nearest minute. The programme included advertisements, which lasted 3.5 minutes, correct to one decimal place. What are the limits of the time of the programme which was devoted to news?

7 A bus will leave the depot at 8.10, to the nearest minute. I leave home at 7.50, and my walk to the depot takes 16 minutes, to the nearest 2 minutes. What are the limits between which my waiting time lies?

8 A train is scheduled to leave a station at 12.35, correct to the nearest minute. My walk to the station takes 8 minutes, correct to the nearest 2 minutes. When should I leave home, to be sure of getting to the station before the train leaves?

9 A film will start at 8.10, correct to the nearest 5 minutes. The length of the film is 93 minutes, correct to the nearest minute. What is the earliest that the film will end?

10 A builder tells his client that he will be able to start work in 20 days, give or take 5 days, and that the job will take 15 days, give or take 2 days. What is the latest that the job will be finished by?

11 The area of a floor is 23 m², correct to the nearest 1 m². The area of a rug is 6.5 m², correct to the nearest 0.5 m². What is the least possible area of floor not covered by the rug?

12 I started the day with £60, correct to the nearest £10. I spent £25, correct to the nearest £5. What is the least amount I have left?

Multiplication and division

Suppose two inaccurate quantities are multiplied.

  The greatest possible value of the product is found by multiplying the greatest values of the original quantities.
  The least possible value of the product is found by multiplying the least values of the original quantities.

Suppose one inaccurate quantity is divided by another.

  The greatest possible value of the result is found by dividing the greatest value of the first quantity by the *least* value of the second.
  The least possible value of the result is found by dividing the least value of the first quantity by the *greatest* value of the second.

Examples   A rectangular field is 80 m long and 60 m wide, both figures being given to the nearest 10 m. What is the least possible value of the area?

The least values of the length and width are 75 m and 55 m respectively. Multiply these to find the least area.

$$75 \times 55 = 4125$$

The least area is 4125 m².

A distance is given as 100 km, to the nearest 10 km. A car is driven at 80 km per hour, to the nearest 10 km per hour. What is the least possible time for the journey?

Take the least possible value for the distance, 95 km, and divide that by the greatest possible value for the speed, 85 km per hour.

$$95 \div 85 = 1.12$$

The least possible time is 1.12 hours.

Exercise 12.4

1 A rectangle is measured as 20 cm by 25 cm, both figures being given to the nearest centimetre. Find the limits between which the area of the rectangle lies.
2 The radius of a circle is 3.4 cm, correct to one decimal place. Find the limits between which the area of the circle lies.
3 The mass of a lump of metal is 1200 kg, correct to the nearest 100 kg, and its volume is 0.15 m³, correct to two decimal places. Find the limits between which the density of the metal lies.

Remember:

Density is mass divided by volume.

4 A sprinter runs for 10.3 seconds at a speed of 7.9 m/s. Both figures are given correct to one decimal place. Find the limits between which the distance run lies.
5 A city has 560 000 inhabitants, correct to the nearest 10 000, and its area is 23 km², correct to the nearest 1 km². Find the limits between which the population density lies.
6 An electrical heater has power 2.1 kW. It is kept on for 4.5 hours. Both figures are correct to one decimal place. Find the limits between which the total energy consumption (in kW hours) lies.
7 The electrical heater of question 6 uses up 9.7 kW hours of energy, correct to one decimal place. Find the least possible time it could have been on.
8 A company employs 2600 people, correct to the nearest 100. Their average wage is £18 000, correct to the nearest £1000. Find the least possible value of the total wage bill of the company.
9 The area of a triangle is 27 cm², and its base is 6 cm. Both figures are given to the nearest whole number. Find the limits between which the height of the triangle lies.
10 The area of a rectangle is 120 m², correct to the nearest 10 m², and its length is 14 m, correct to the nearest metre. Find the limits between which the width of the rectangle lies.
11 The volume of a cuboid is 650 cm³, correct to the nearest 10 cm³, and its length and breadth are 12 cm and 5 cm respectively, both correct to the nearest centimetre. Find the limits between which the height of the cuboid lies.
12 The volume of a can of soup is 820 cm³, correct to the nearest 10 cm³. Its radius is 5.1 cm, correct to one decimal place. What is the greatest possible height of the can?
13 The volume of a petrol can is 5600 cm³, correct to the nearest 100 cm³. Its height is 14 cm, correct to the nearest centimetre. What is the least possible radius of the can?
14 In a market in France you see tomatoes on sale at 8 francs per *livre*. One *livre* is 0.45 kg, correct to two decimal places, and there are 11 francs per £, correct to the nearest franc. Find the greatest possible equivalent price of the tomatoes in £ per kilogram.

SUMMARY

- The **error** in a measurement is the difference between the measurement and the true value. For example, if a length of 10 m is measured as 9 m or 11 m, the error is 1 m. The **relative error** is the error divided by the true value. If we do not know the true value, divide by the measured value. When the relative error is expressed as a percentage, this is called the **percentage error**. In the measurement above, the relative error is $\frac{1}{10}$, and the percentage error is 10%.
- If a quantity is given correct to a certain unit, the error is up to half that unit. For example, if a length is given as 1.32 km, correct to two decimal places, the maximum error is 0.005 km.
- When measurements are added or subtracted, the errors are added.
- When measurements are multiplied, the greatest possible value of the product is found by multiplying the greatest possible values of the original measurements.
- When one measurement is divided by another, the greatest possible value is found by dividing the greatest possible value of the first measurement by the least possible value of the second.

Exercise 12A

1 A circle has radius exactly 4 cm. Its circumference is found, using the value of $\frac{22}{7}$ for π. Use your calculator to find the error in the calculation.

2 Find the relative error in the calculation of question 1.

3 Find the percentage error in the calculation of question 1.

4 A temperature is given as 18.7 °C, correct to one decimal place. Give the limits between which the temperature lies.

5 On a copy of this number line, indicate the possible values of the temperature of question 4.

6 A weight was measured. The possible values are shown on this number line. To what accuracy was the weight measured?

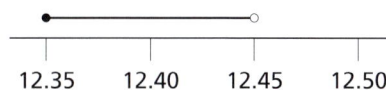

7 A cup of white coffee contains 110 ml of coffee, to the nearest 10 ml, and 15 ml of milk, to the nearest 5 ml. Find the limits between which the total volume lies.

8 At the beginning of the month I had £850 in my bank account, correct to the nearest £10. Over the month no money went into the account, and the amount that went out was £550, correct to the nearest £50. What is the least amount that can now be in my account?

9 The side of a cube is 3.4 cm, correct to one decimal place. What is the upper limit on the volume of the cube?

10 For 12 gallons of petrol (correct to the nearest gallon) my car travelled 340 miles (correct to the nearest 10 miles). What is the least possible value of the fuel economy of my car, in miles per gallon?

Exercise 12B

1 The true weight of a parcel is 1.7 kg, but the scales at the post office measure it as 1.9 kg. What is the error in the measurement?

2 What is the relative error in the measurement of question 1?

3 What is the percentage error in the measurement of question 1?

4 The area of a plot of land is given as 1200 m², correct to the nearest 100 m². What are the limits between which the true area lies?

5 Copy the number line below and on it indicate the limits between which the area of question 4 lies.

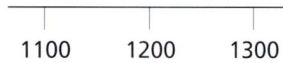

```
      ┬───────┬───────┬
    1100    1200    1300
```

6 The pressure in a car tyre was measured. The range of possible values is shown on the number line. What was the accuracy of the measurement?

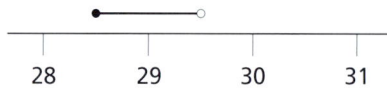

```
       ●───────○
    ┬───────┬───────┬───────┬
   28      29      30      31
```

7 The annual incomes of Mr and Mrs Smith are £19 000 and £17 000 respectively, both figures being given correct to the nearest £1000. What are the limits between which their total income lies?

8 The weight of a full bottle is 1040 grams, correct to the nearest 10 grams, and the weight of the glass is 275 grams, correct to the nearest 5 grams. What is the least possible weight of the liquid in the bottle?

9 The fuel economy of a car is 12 litres per 100 km, correct to the nearest litre. The car is driven for 240 km, correct to the nearest 10 km. What is the upper limit on the amount of fuel used?

10 The strength of an alcoholic beverage is measured by its **alcohol by volume** (a.b.v.). This is the percentage of the liquid which is alcohol. Suppose 2.4 litres of beer (correct to one decimal place) contains 0.11 litres of alcohol (correct to two decimal places). What is the least possible a.b.v. of the beer?

Exercise 12C (Ma1)

How good are you at estimating distances, angles, weights and so on? You can test in two ways, as follows.

1 a What is the length of this line? Make your guess, and then measure with a ruler. What was the error and the relative error?

```
    ─────────────────────────
```

  b Without a ruler, draw a line which you think is 5 cm long. Measure it, and find the error and the relative error.

2 a What is this angle? Make your guess, and then measure with a protractor. What was the error and the relative error?

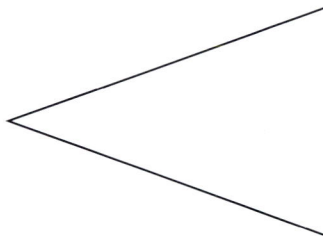

  b Without a protractor, draw an angle which you think is 50°. Measure it, and find the error and the relative error.

3 Devise experiments to see how good you are at estimating weights or times.

Exercise 12D   Ma1

When measurements are added or subtracted, the errors are added. What happens when the measurements are multiplied or divided?

1 Suppose the sides of a rectangle are 6.0 cm and 8.0 cm, both figures being correct to one decimal place. For the lengths and for the area of the rectangle, find the errors and the relative errors. Can you find a rule? Test with another pair of numbers.

2 Can you show why your rule holds?

3 Suppose we find $x \div y$, where $x = 68$ and $y = 80$, both numbers being given correct to the nearest whole number. For x, y and $x \div y$, find the errors and the relative errors. Can you find a rule? Test with another pair of numbers.

4 Can you show why your rule holds?

Exercise 12E

To find the number π you can use the button on your calculator. How is π found in the first place? There are expressions which get closer and closer to π, without ever reaching it. This exercise uses a spreadsheet to evaluate these expressions.

1 One result is that $\dfrac{\pi^2}{6} = \dfrac{1}{1^2} + \dfrac{1}{2^2} + \dfrac{1}{3^2} + \dots$. In the A column enter 1, 2, 3,... up to 200. You can use

the 'fill' command for this, or enter 1 in A1 and =A1+1 in A2, then copy.
The B column will contain the sums of the series. Enter 1 in B1, then enter =B1+1/A2^2 in B2. Copy this formula down to B200.

How close is B200 to $\dfrac{\pi^2}{6}$? You could use the C column to show the error.

After how many terms do we get $\dfrac{\pi^2}{6}$ correct to three decimal places?

	File Edit View Insert Format Tools Data Window Help

	A	B	C	D	E	F	G
1	1	1					
2	2	1.25					
3	3						
4	4						

B2 = =B1+1/A2^2

2 Another result is that $\frac{1}{4}\pi = 1 - \frac{1}{3} + \frac{1}{5} - \frac{1}{7} + \dots$. Use the F column for 1, −3, +5, −7 and so on. In F1 enter =−(2*A1−1)*(−1)^A1. Copy this down the F column. Put 1 in G1, then =G1+1/F2 in G2. Copy down the G column to G200.

How close is G200 to $\frac{1}{4}\pi$? After how many terms do we get $\frac{1}{4}\pi$ correct to three decimal places?

3 The expressions above are inefficient methods of finding π. An expression which reaches π quickly, and which has been used in the past to evaluate π, is

$$\pi = 16\left(\tfrac{1}{5} - \tfrac{1}{3}\left(\tfrac{1}{5}\right)^3 + \tfrac{1}{5}\left(\tfrac{1}{5}\right)^5 - \tfrac{1}{7}\left(\tfrac{1}{7}\right)^7 + \dots\right) - 4\left(\tfrac{1}{239} - \tfrac{1}{3}\left(\tfrac{1}{239}\right)^3 + \tfrac{1}{5}\left(\tfrac{1}{239}\right)^5 - \tfrac{1}{7}\left(\tfrac{1}{239}\right)^7 + \dots\right)$$

Adapt your spreadsheet for this expression, and hence find π to eight decimal places.

13 Vectors

Translation and vectors

A **translation** moves points in a particular direction, without altering the shape or size of figures. A translation can be given by a **vector** $\begin{pmatrix} x \\ y \end{pmatrix}$, which shows how much the translation moves to the right and how much upwards. The top number shows the change in the x-coordinate of points, and the bottom number shows the change in the y-coordinate. For example

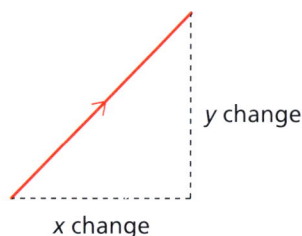

y change

x change

- Use negative numbers for motion to the left or downwards.
- The right–left movement (the change in x-coordinate) is on the top of the vector, and the up–down movement (the change in y-coordinate) is on the bottom.

$\begin{pmatrix} 4 \\ 5 \end{pmatrix}$ moves points 4 units to the right and 5 units upwards.

$\begin{pmatrix} -2 \\ -3 \end{pmatrix}$ moves points 2 units to the left and 3 units downwards.

The numbers occurring in a vector are its **components**. For the vector $\begin{pmatrix} 2 \\ 4 \end{pmatrix}$, the x-component is 2 and the y-component is 4.

Examples A translation moves (3, 7) to (5, 2). Write down the vector for this translation.

The value of x has increased by 2, and the value of y has decreased by 5.

The vector is $\begin{pmatrix} 2 \\ -5 \end{pmatrix}$.

A translation has vector $\begin{pmatrix} -3 \\ 4 \end{pmatrix}$. Where does the translation take (1, 7)?

Subtract 3 from the x-coordinate, and add 4 to the y-coordinate.
The point is taken to $(-2, 11)$.

Exercise 13.1

1 A translation takes $(1, -4)$ to $(4, -7)$. Write down the vector for this translation.
2 A translation takes $(4, 2)$ to $(-2, 9)$. Write down the vector for this translation.
3 Write down the vectors which take
   a $(1, 3)$ to $(5, 0)$ **b** $(-2, 8)$ to $(7, 3)$
   c $(4, -2)$ to $(1, -7)$ **d** $(-1, -3)$ to $(-5, -6)$

4 **a** Describe the action of the translation whose vector is $\begin{pmatrix} 3 \\ -2 \end{pmatrix}$.

  b Where does this translation take $(4, -2)$?

5 **a** Describe the action of the translation whose vector is $\begin{pmatrix} -4 \\ 3 \end{pmatrix}$.

  b Where does this translation take $(1, 2)$?

6 Where does the vector $\begin{pmatrix} -5 \\ 3 \end{pmatrix}$ take these points?

  a $(2, 6)$ **b** $(-2, 7)$
  c $(1, -9)$ **d** $(-2, -7)$

Working with vectors

Vectors have many other uses besides representing translations. As we shall see later in the chapter, they are used for force and velocity, and many other quantities. The general definition is as follows.

> magnitude = size

● *A vector is a quantity which has direction as well as magnitude.*

For example, for a translation you need to know in which direction points are being moved, as well as how far they are being moved. The magnitude of a translation vector is the distance through which it moves points. The direction gives which way the points are moving.

A quantity which has magnitude only is a **scalar**. The amount of money you have is a scalar. It doesn't make sense to say: 'I have £10 in a northerly direction.'

Exercise 13.2

1 Which of the following are vectors, and which are scalars?
  a mass   **b** temperature
  c electric current   **d** gravitational force
  e volume
2 The *speed* of an object tells us how fast it is going, without any regard to direction. The *velocity* of an object gives its speed and its direction. So speed is not a vector, but velocity is. Which of the following statements refer to speed and which to velocity?
  a Jason can run at 8 metres per second.
  b The plane is flying North East at 800 km per hour.
  c Mandy threw the ball upwards at 10 metres per second.
  d The bullet leaves the rifle at 800 metres per second.
3 A car is being driven at 30 m.p.h. in a circle. Which of these are true?
  a The speed is constant.   **b** The velocity is constant.
4 Which of the following statements refer to vectors and which to scalars?
  a Bristol is 100 miles from London.
  b Sheffield is 100 km East of Birkenhead.
  c The car has a mass of 1100 kg.
  d The weight of the car acts towards the centre of the Earth.

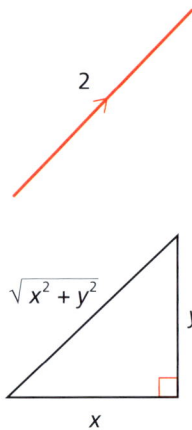

We can illustrate vectors as line segments, with an arrowhead to show their direction. The vector shown has magnitude 2 units, and direction up and to the right.

If you are given a vector in column form, then use Pythagoras' theorem to find its length. The vector $\begin{pmatrix} x \\ y \end{pmatrix}$ represents a translation x to the right and y upwards, so the total distance of the translation is

$$\sqrt{x^2 + y^2}$$

This is the magnitude of the vector $\begin{pmatrix} x \\ y \end{pmatrix}$.

In type or print, vectors are often written with bold letters, for example **a**. When writing by hand, it is easier to write them underlined, as <u>a</u>.

Example   Find the magnitude of the vector $\begin{pmatrix} 5 \\ -12 \end{pmatrix}$.

Using the formula, putting $x = 5$ and $y = -12$

$$\sqrt{5^2 + (-12)^2} = \sqrt{25 + 144}$$
$$= \sqrt{169}$$
$$= 13$$

The magnitude is 13.

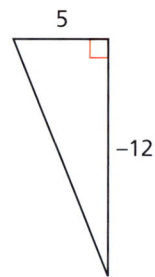

Exercise 13.3

You will need:
- squared paper

1 Write down in column form the vectors shown in the diagram below.

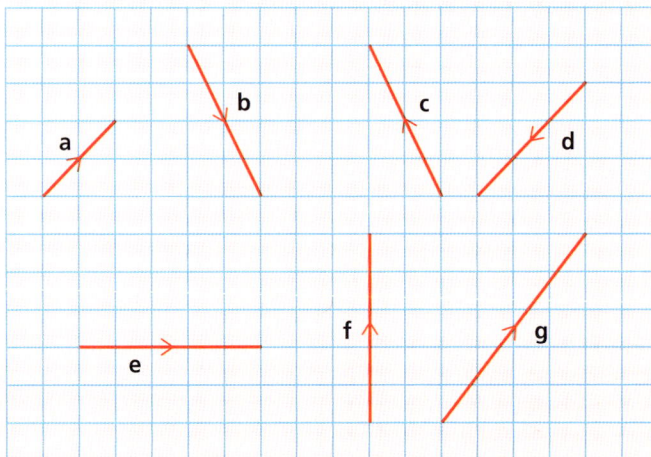

2 On squared paper draw vectors for each of the following.

$$\begin{pmatrix} 2 \\ 4 \end{pmatrix} \begin{pmatrix} 3 \\ -1 \end{pmatrix} \begin{pmatrix} -3 \\ 4 \end{pmatrix} \begin{pmatrix} -3 \\ -2 \end{pmatrix} \begin{pmatrix} 3 \\ 0 \end{pmatrix} \begin{pmatrix} 0 \\ -4 \end{pmatrix}$$

3 Find the magnitudes of the vectors in question 1.

It is important to realise that, although vectors have direction and magnitude, they do not have position. A vector can start from anywhere. In the diagram below, all the vectors have the same magnitude and direction, though they have different starting points. They are all the same vector.

This point is emphasised in the following example.

Example   A translation has vector $\begin{pmatrix} 3 \\ 2 \end{pmatrix}$. Find where the translation takes $(4, 1)$ and $(-2, 3)$. Illustrate these moves on a diagram.

The translation moves points 3 units to the right and 2 upwards. In other words, it adds 3 to the x-coordinate and 2 to the y-coordinate.

$(4, 1)$ is moved to $(7, 3)$
$(-2, 3)$ is moved to $(1, 5)$

The translation is shown in the diagram. Notice that the vector takes $(4, 1)$ to $(7, 3)$, and it takes $(-2, 3)$ to $(1, 5)$. It is the same vector, even though the starting point varies. Vectors do not have position.

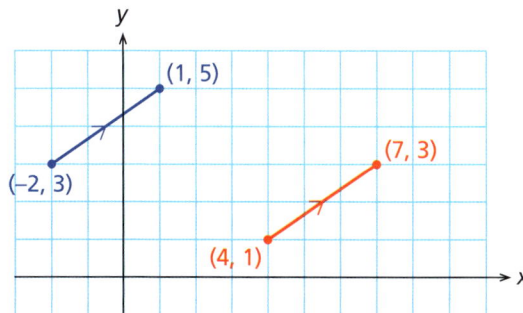

Exercise 13.4

You will need:
• squared paper

1 A translation has vector $\begin{pmatrix} 3 \\ -4 \end{pmatrix}$. Find where the translation takes $(2, 7)$

and $(-2, 3)$. Illustrate the moves on a diagram.

2 A translation has vector $\begin{pmatrix} -5 \\ 2 \end{pmatrix}$. Find where the translation takes $(-2, 1)$ and $(4, -1)$. Illustrate the

moves on a diagram.

3 A translation moves $(3, 4)$ to $(8, 1)$. Where does it take $(-2, 2)$?
4 A translation moves $(1, -2)$ to $(-3, 7)$. Where does it take $(5, -3)$?
5 A translation moves $(2, 5)$ to $(1, 8)$. Where does it take (x, y)?
6 A translation moves (x, y) to $(x + 7, y - 3)$. Write down the vector for this translation.

So vectors can be shown in two ways: as a column vector of two numbers, or as a directed line segment.

Addition of vectors

Suppose we do one translation after another. The combined translation is found by finding the total x-movement and the total y-movement. So add the two x-movements and add the two y-movements. In terms of vectors, add the top terms and add the bottom terms.

For example, suppose the two translations are represented by $\binom{6}{7}$ and $\binom{2}{1}$. Then the combined translation is represented by

$$\binom{6}{7} + \binom{2}{1} = \binom{8}{8}$$

● *To add vectors, add the corresponding components.*

If the vectors are represented by line segments, then add them by combining them as a triangle as shown. In the diagram **a** + **b** is found by putting the tail of **b** onto the head of **a**, and then drawing the third side of the triangle. The result is called a **triangle of vectors**.

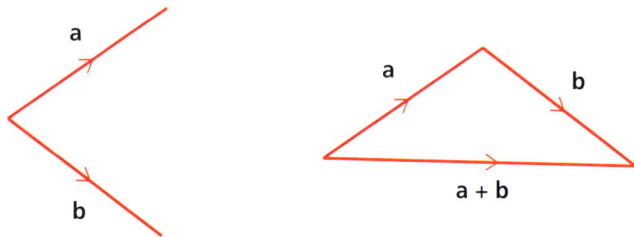

We would get exactly the same result by putting the tail of **a** onto the head of **b**. This diagram shows that the results are the same. It is called a **parallelogram of vectors**.

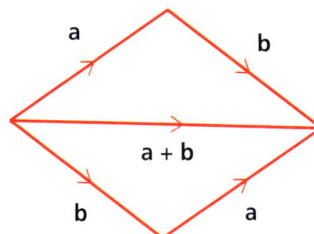

Example   The vectors **a** and **b** are shown. Draw the vector **a** + **b**.

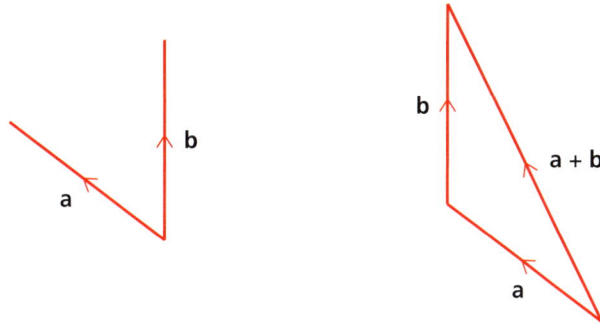

Put the tail of **b** onto the head of **a**, then join the tail of **a** to the head of **b**. The result is **a** + **b**.

Exercise 13.5

You will need:
• squared paper

1 Make a copy of the vectors **a** and **b** shown. Construct **a** + **b**.

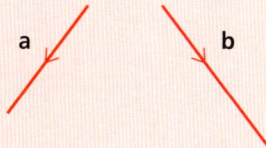

2 Make a copy of the vectors **u** and **v** shown. Construct **u** + **v**.

3 Let $\mathbf{a} = \begin{pmatrix} 2 \\ 4 \end{pmatrix}$ and $\mathbf{b} = \begin{pmatrix} 4 \\ -3 \end{pmatrix}$. Find **a** + **b**.

4 On squared paper, draw the vectors **a** and **b** of question 3. Construct the vector **a** + **b**. Check that the result agrees with your answer for question 3.

5 With **a** and **b** as in question 3, find the magnitudes of **a**, **b** and **a** + **b**.

6 Let $\mathbf{u} = \begin{pmatrix} 6 \\ -1 \end{pmatrix}$ and $\mathbf{v} = \begin{pmatrix} -2 \\ 5 \end{pmatrix}$.

i)   Find **u** + **v**.
ii)   On squared paper, draw the vectors **u**, **v** and **u** + **v**. Check that the result agrees with your answer for part i).
iii) Find the magnitudes of **u**, **v** and **u** + **v**.

Multiplying vectors by scalars

A scalar is an ordinary number. To multiply a vector by a scalar, just multiply all the components by the scalar.

$$3 \times \begin{pmatrix} 1 \\ -2 \end{pmatrix} = \begin{pmatrix} 3 \\ -6 \end{pmatrix}$$

Geometrically, multiply a vector by a scalar to alter its length. The diagram below shows $\begin{pmatrix} 1 \\ -2 \end{pmatrix}$ and $3 \times \begin{pmatrix} 1 \\ -2 \end{pmatrix}$. The second vector is 3 times as long as the first vector. Notice that the two vectors are parallel, so the direction has not changed.

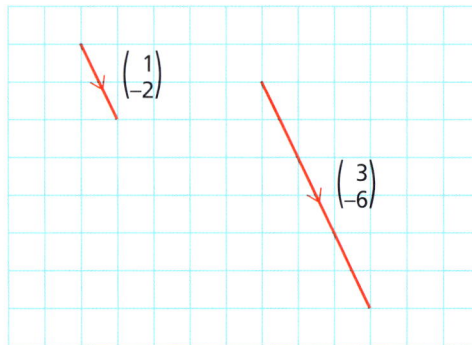

If the scalar multiplier is negative, then the direction of the vector is reversed.

$$-2 \times \begin{pmatrix} -3 \\ 4 \end{pmatrix} = \begin{pmatrix} 6 \\ -8 \end{pmatrix}$$

In the diagram, $\begin{pmatrix} 6 \\ -8 \end{pmatrix}$ is twice as long as $\begin{pmatrix} -3 \\ 4 \end{pmatrix}$, and in the reverse direction.

The directions are still parallel.

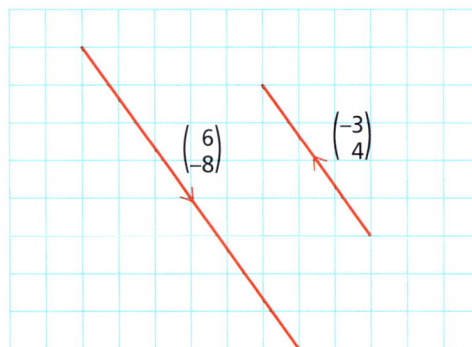

In particular, $-1 \times \mathbf{a}$ just reverses the direction of $\mathbf{a}$, without altering its size. This enables us to *subtract* vectors. To subtract $\mathbf{a}$, add $-\mathbf{a}$.

$$\mathbf{b} - \mathbf{a} = \mathbf{b} + (-\mathbf{a})$$

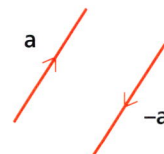

Examples   The vectors shown are **a** and **b**. Draw

i) **a** − **b**   ii) 2**a** + 3**b**

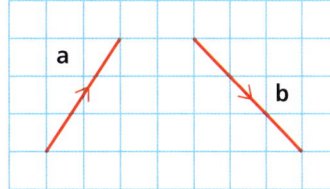

i) Reverse the direction of **b**, and then add it to **a** by the method on page 186.
ii) Double the length of **a**, and triple the length of **b**. Add them by the method
    on page 186.

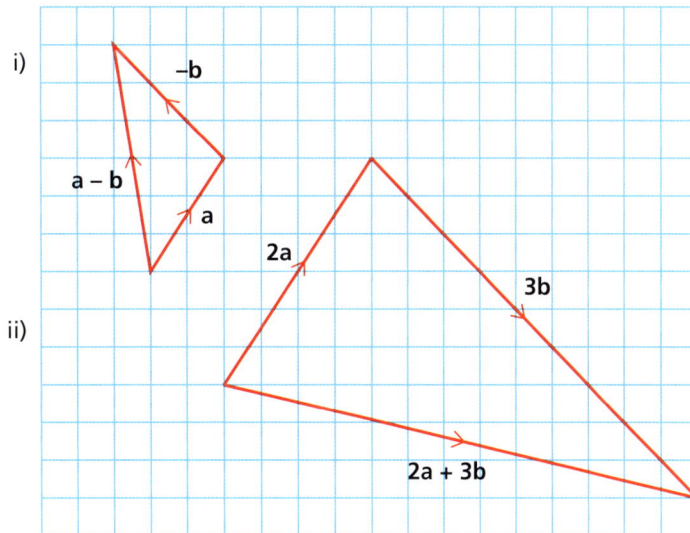

Let $\mathbf{a} = \begin{pmatrix} 5 \\ 2 \end{pmatrix}$ and $\mathbf{b} = \begin{pmatrix} -2 \\ 4 \end{pmatrix}$.

Find i) **a** − **b**   ii) 2**a** + 3**b**

i) Subtract the coordinates.

$$\mathbf{a} - \mathbf{b} = \begin{pmatrix} 5 \\ 2 \end{pmatrix} - \begin{pmatrix} -2 \\ 4 \end{pmatrix} = \begin{pmatrix} 5 - (-2) \\ -2 \end{pmatrix}$$

$$\mathbf{a} - \mathbf{b} = \begin{pmatrix} 7 \\ -2 \end{pmatrix}$$

ii) $2\mathbf{a} + 3\mathbf{b} = 2 \times \begin{pmatrix} 5 \\ 2 \end{pmatrix} + 3 \times \begin{pmatrix} -2 \\ 4 \end{pmatrix} = \begin{pmatrix} 10 \\ 4 \end{pmatrix} + \begin{pmatrix} -6 \\ 12 \end{pmatrix}$

$$2\mathbf{a} + 3\mathbf{b} = \begin{pmatrix} 4 \\ 16 \end{pmatrix}$$

Remember:

To solve simultaneous equations, adjust them so that the x-terms or the y-terms are the same size. Then add if they have different signs, and subtract if they have same signs. This eliminates one of the unknowns.

With **a** and **b** as in the example on page 189, find x and y such that $x\mathbf{a} + y\mathbf{b} = \begin{pmatrix} 13 \\ 10 \end{pmatrix}$.

We have $x\begin{pmatrix} 5 \\ 2 \end{pmatrix} + y\begin{pmatrix} -2 \\ 4 \end{pmatrix} = \begin{pmatrix} 13 \\ 10 \end{pmatrix}$. This comes to

$$5x - 2y = 13$$
$$2x + 4y = 10$$

This gives a pair of simultaneous equations.

$$5x - 2y = 13 \qquad [1]$$
$$2x + 4y = 10 \qquad [2]$$

To eliminate the y-terms, halve [2] and then add to [1].

$$x + 2y = 5 \qquad [3] = \tfrac{1}{2} \times [2]$$
$$6x \quad = 18 \qquad [3] + [1]$$

Hence $x = 3$. Substitute in [2] to get

$$6 + 4y = 10 \qquad \text{hence } y = 1$$

So $x = 3$ and $y = 1$.

   Check. Substitute these values into [1] to check the answer.

$$5 \times 3 - 2 \times 1 = 15 - 2 = 13$$

The answer is correct.

Exercise 13.6

1 Make a copy of the vector **a**, and then draw
   i) 2**a**   ii) −**a**   iii) −3**a**

2 Make a copy of the vector **u**, and then draw
   i) 3**u**   ii) −**u**   iii) $\frac{1}{2}$**u**

3 Let $\mathbf{a} = \begin{pmatrix} 4 \\ -3 \end{pmatrix}$. Evaluate

   i) −**a**   ii) 3**a**   iii) −2**a**

4 Let $\mathbf{u}=\begin{pmatrix}-5\\3\end{pmatrix}$. Evaluate

  i) $-\mathbf{u}$
  ii) $\frac{1}{2}\mathbf{u}$
  iii) $-\frac{1}{4}\mathbf{u}$

5 Make a copy of the vectors
u and **v** in the diagram, and
construct the vectors
  i) $\mathbf{u}-\mathbf{v}$
  ii) $3\mathbf{u}+\mathbf{v}$
  iii) $2\mathbf{u}-3\mathbf{v}$

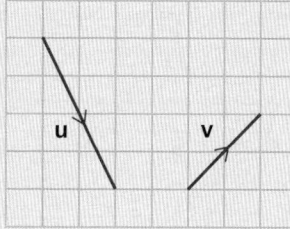

6 Let $\mathbf{u}=\begin{pmatrix}4\\-7\end{pmatrix}$ and $\mathbf{v}=\begin{pmatrix}6\\3\end{pmatrix}$. Evaluate

  i) $\mathbf{u}-\mathbf{v}$
  ii) $2\mathbf{u}+4\mathbf{v}$
  iii) $-3\mathbf{u}+4\mathbf{v}$
  iv) $4\mathbf{u}-\frac{1}{2}\mathbf{v}$

7 With **u** and **v** as in question 6, solve the equation $x\mathbf{u}+y\mathbf{v}=\begin{pmatrix}12\\33\end{pmatrix}$.

8 Let $\mathbf{a}=\begin{pmatrix}-5\\3\end{pmatrix}$ and $\mathbf{b}=\begin{pmatrix}8\\-6\end{pmatrix}$. Evaluate

  i) $\mathbf{a}-\mathbf{b}$
  ii) $\mathbf{b}-\mathbf{a}$
  iii) $3\mathbf{a}+7\mathbf{b}$
  iv) $4\mathbf{a}-3\mathbf{b}$

9 With **a** and **b** as in question 8, solve the equation $x\mathbf{a}+y\mathbf{b}=\begin{pmatrix}1\\-3\end{pmatrix}$.

10 Solve these equations.

a $\begin{pmatrix}5\\y\end{pmatrix}+\begin{pmatrix}x\\7\end{pmatrix}=\begin{pmatrix}8\\-2\end{pmatrix}$ **b** $\begin{pmatrix}2k\\7\end{pmatrix}+\begin{pmatrix}7\\3m\end{pmatrix}=\begin{pmatrix}11\\4\end{pmatrix}$

c $\begin{pmatrix}3\\k\end{pmatrix}+\begin{pmatrix}3m\\2\end{pmatrix}=\begin{pmatrix}m\\4\end{pmatrix}$ **d** $\begin{pmatrix}6\\y\end{pmatrix}+\begin{pmatrix}x\\7\end{pmatrix}=\begin{pmatrix}3x\\x\end{pmatrix}$

e $\begin{pmatrix}a\\b\end{pmatrix}+\begin{pmatrix}b\\-a\end{pmatrix}=\begin{pmatrix}7\\-3\end{pmatrix}$ **f** $x\begin{pmatrix}1\\5\end{pmatrix}+y\begin{pmatrix}2\\3\end{pmatrix}=\begin{pmatrix}9\\31\end{pmatrix}$

g $p\begin{pmatrix}3\\4\end{pmatrix}+q\begin{pmatrix}-2\\6\end{pmatrix}=\begin{pmatrix}19\\8\end{pmatrix}$ **h** $k\begin{pmatrix}4\\3\end{pmatrix}+m\begin{pmatrix}-3\\-2\end{pmatrix}=\begin{pmatrix}18\\14\end{pmatrix}$

Use of vectors

Force

Vectors are used throughout science to represent any quantity for which the direction is important. When designing a bridge, an engineer needs to know the directions of the forces as well as their magnitudes. A meteorologist needs to know the direction of the wind as well as its speed. So both force and velocity are represented by vectors.

If two forces act on a body, the total force is found by adding their vectors.

Example A force of 8 newtons acts due North on a body. What extra force should act on the body so that the total force is 6 newtons acting due East?

> 1 newton (N) = the force required to produce an acceleration of 1 metre per second squared in a mass of 1 kilogram.

Let the original force be **F** and the extra force be **G**. Let the total force, of 6 newtons acting East, be **H**. Then

$$\mathbf{F} + \mathbf{G} = \mathbf{H}$$

Hence $\mathbf{G} = \mathbf{H} - \mathbf{F}$. The diagram shows $-\mathbf{F}$ added to **H**. The third side of the triangle gives **G**.

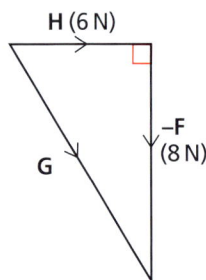

> **Remember:**
>
> Pythagoras' theorem
> $c^2 = a^2 + b^2$

Find the magnitude of **G** by Pythagoras' theorem, and its direction by trigonometry.

The magnitude of **G** is $\sqrt{6^2 + 8^2} = \sqrt{100} = 10$ newtons.
The direction that **G** makes with East is given by $\tan^{-1}\left(\frac{8}{6}\right) = 53°$.

> **Remember:**
>
> SOHCAHTOA

The extra force is 10 newtons acting on a bearing of 143°.

Exercise 13.7

1 A force of 2 N acts due North, and a force of 3 N acts due East. Find the magnitude and direction of the combined force.
2 A force of 20 N acts North East (i.e. on a bearing of 045°) and a force of 30 N acts North West (on a bearing of 315°). Find the magnitude and direction of the combined force.
3 Forces of 40 N and 50 N act on bearings of 045° and 135° respectively. Find the magnitude and direction of the combined force.
4 A force of 10 N acts due South. What Eastwards force should be applied to ensure that the total force acts South East? What will be the magnitude of this total force?
5 A force of 30 N acts North East. What North West force will ensure that the combined force acts North? What will be the magnitude of this total force?

6 A force of 20 N acts due North. What extra force should be applied to ensure that the total force is 15 N due West?

7 A force of 10 N acts due North. In what direction should an extra force of 15 N be applied to ensure that the total force acts due East?

Velocity

Suppose you are swimming in a river, which itself is moving. Then your total velocity is found by adding the velocity of the river to your velocity relative to the river.

Example A river is flowing at 2 m/s. You can swim at 3 m/s. You are on one bank of the river and want to reach a point which is directly opposite you, 100 m away on the other bank of the river. In which direction should you swim, and how long will it take you?

Let the velocity of the current be **v**, let your velocity be **u** and let the total velocity be **w**.

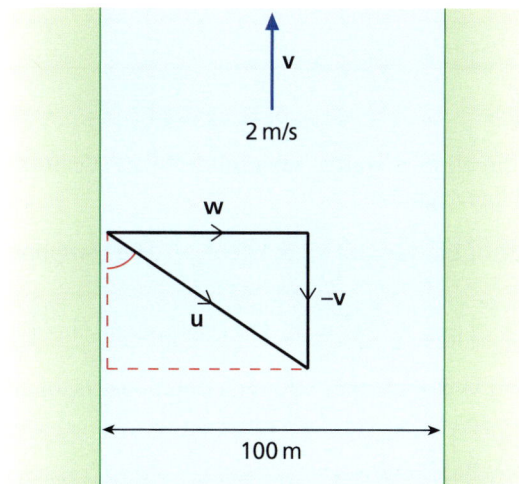

Then

$$\mathbf{v} + \mathbf{u} = \mathbf{w}$$

Hence $\mathbf{u} = \mathbf{w} - \mathbf{v}$.

We need to ensure that **w** is directly across the river (at right angles to the bank). This gives us the angle to the bank at which to swim.

$$\text{angle} = \cos^{-1}\left(\tfrac{2}{3}\right)$$

You should swim at 48° to the bank.

Your total speed is the magnitude of **w**. Find this by Pythagoras' theorem.

$$\sqrt{3^2 - 2^2} = \sqrt{5}$$

So the speed is about 2.24 m/s. Divide this into 100.

It will take 44.7 s to cross the river.

Exercise 13.8

1 A river is 100 m wide, and is flowing at 2 km/h. A man swims at 3 km/h, directly across the river (at right angles to the bank). What will be his actual direction and speed, and how far downstream will he be when he reaches the opposite side?

2 There is a wind of 50 km/h blowing South. A plane can fly at 300 km/h. If the pilot sets a course of due West, what will be the actual speed and direction of the plane?

3 The sea current is 1 km/h West. A man swims directly North at 3 km/h. What is his actual direction and speed?

4 A river 200 m wide flows at 1 m/s. A man can row at 3 m/s. He wants to cross to the point directly opposite. In what direction should he row, and how long will it take him?

5 There is a wind of 40 km/h blowing North. A light aircraft can fly at 200 km/h, and its pilot wants to travel to a place 400 km due East. On what bearing should the plane be headed, and how long will the journey take?

6 The sea current is 1 km/h West. A woman who can swim at 3 km/h wants to swim North. In which direction should she swim, and how long will it take her to cover 100 m?

7 A gazelle is running due North at 40 km/h. A cheetah, which can run at 60 km/h, spots the gazelle when it is 100 m due East and sets off to intercept. Assuming that the gazelle does not notice its danger, and so does not change its velocity, find the time taken for the cheetah to catch the gazelle.

SUMMARY

■ A **vector** has direction as well as **magnitude** (size). Vectors can be represented as columns of numbers or as directed line segments.

■ The magnitude of the vector $\begin{pmatrix} x \\ y \end{pmatrix}$ is $\sqrt{x^2 + y^2}$. The **components** are x and y. For example, the magnitude of $\begin{pmatrix} 8 \\ 6 \end{pmatrix}$ is $\sqrt{8^2 + 6^2} = 10$. The components are 8 and 6.

■ To add together vectors, either add their components, or complete a **triangle of vectors**. For example, $\begin{pmatrix} 8 \\ 6 \end{pmatrix} + \begin{pmatrix} 7 \\ -2 \end{pmatrix} = \begin{pmatrix} 15 \\ 4 \end{pmatrix}$.

■ To multiply a vector by a **scalar** (a number), either multiply its components by the scalar, or multiply its length by the scalar if you are drawing it as a directed line segment. For example, $3 \times \begin{pmatrix} 8 \\ 6 \end{pmatrix} = \begin{pmatrix} 24 \\ 18 \end{pmatrix}$.

■ Vectors can be used to represent forces or velocities.

Exercise 13A

1 Write down the vector for the translation which takes $(3, 1)$ to $(-1, 5)$.

2 Do the following statements refer to velocity or speed?
   a The world record for a car on land is 766 m.p.h.
   b The Shinkansen train travels from Tokyo to Osaka at 270 km/h.

3 The diagram shows **a** and **b**. On a copy of the diagram illustrate **a** + **b**.

4 Find the magnitude of the vector $\begin{pmatrix} 4 \\ -7 \end{pmatrix}$.

5 Find the value of x if the magnitude of $\begin{pmatrix} x \\ 3 \end{pmatrix}$ is 14.

6 Let $\mathbf{a} = \begin{pmatrix} 2 \\ -3 \end{pmatrix}$ and $\mathbf{b} = \begin{pmatrix} -3 \\ 2 \end{pmatrix}$. Find $2\mathbf{a} + 3\mathbf{b}$.

7 Solve the equation $\begin{pmatrix} x \\ 2y \end{pmatrix} + \begin{pmatrix} 3y \\ -x \end{pmatrix} = \begin{pmatrix} 11 \\ -1 \end{pmatrix}$.

8 With $\mathbf{a}$ and $\mathbf{b}$ as in question 6, solve the equation $x\mathbf{a} + y\mathbf{b} = \begin{pmatrix} 8 \\ -7 \end{pmatrix}$.

9 A force of 8 N acts due North. What force acting South East will ensure that the total force acts East? What is the magnitude of this total force?

10 A river flows at 2 m/s. A boat which can be rowed at 4 m/s is steered directly across the river. At what speed is the boat travelling, and at what angle to the bank?

Exercise 13B

1 Describe the translation whose vector is $\begin{pmatrix} -2 \\ -5 \end{pmatrix}$.

2 Which of the following is a scalar quantity and which a vector?
  a volume   **b** acceleration
3 On a copy of the diagram, draw the vector $\mathbf{u} - \mathbf{v}$.

4 Find the magnitude of $\begin{pmatrix} 7 \\ -3 \end{pmatrix}$.

5 Find x so that the magnitude of $\begin{pmatrix} x \\ 2x \end{pmatrix}$ is 17.

6 Let $\mathbf{u} = \begin{pmatrix} 8 \\ 3 \end{pmatrix}$ and $\mathbf{v} = \begin{pmatrix} -1 \\ 6 \end{pmatrix}$. Find $3\mathbf{u} - 2\mathbf{v}$.

7 Find x and y, if $\begin{pmatrix} 2x + 3 \\ 4y - 7 \end{pmatrix} = \begin{pmatrix} 11 \\ 11 \end{pmatrix}$.

8 With $\mathbf{u}$ and $\mathbf{v}$ as in question 6, solve the equation $x\mathbf{u} + y\mathbf{v} = \begin{pmatrix} 18 \\ -6 \end{pmatrix}$.

9 A force of 25 N acts North. What extra force of 30 N will ensure that the total force acts East? What is the magnitude of this total force?

10 The wind is blowing due North at 50 km/h. A plane can fly at 250 km/h. If the pilot wants to travel 400 km due West, in what direction should the plane be steered, and how long will the journey take?

Exercise 13C (Ma1)

The **triangle inequality** states that the magnitude of **a** + **b** is at most the sum of the magnitudes of **a** and **b**.

1 Show, by a sketch, why the inequality is true.

2 Verify the inequality for $\mathbf{a} = \begin{pmatrix} 4 \\ 2 \end{pmatrix}$ and $\mathbf{b} = \begin{pmatrix} 2 \\ 7 \end{pmatrix}$.

3 If the magnitude of **a** + **b** is equal to the sum of the magnitudes of **a** and **b**, what can you say about **a** and **b**?

Exercise 13D (Ma1)

Here are some trickier problems about forces and velocity.

1 Weights of 3 kg, 5 kg and 4 kg are on a string, which passes over smooth pegs as shown. What angles do the strings supporting the 5 kg mass make with the vertical?

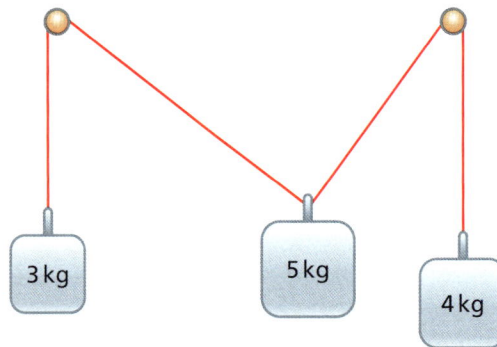

2 A river is 100 m wide, flowing at 4 m per second. You can swim at 3 m per second. Show by construction that you can reach a point which is 100 m downstream. How long will it take you?

3 An antelope is running North at 15 m/s. A rather slow lion spots it when it is 100 m South East, and sets off in pursuit at 10 m/s. Show that the lion will not catch the antelope. If the lion goes due East, what is the closest it will get?

Exercise 13E

A computing language like *Logo* can draw vectors on the computer screen. The command fd 10 will draw a vector of magnitude 10 units, directly up the screen. The command rt 45 fd 20 will draw a vector of magnitude 20 units at an angle of 45°. You can use *Logo* to check some of the results of this chapter. Look at the example on page 192 about the force acting on a body.

The force **F** has magnitude 8 and acts due north. It is represented by fd 8. The extra force **G** has magnitude 10 and acts $\tan^{-1}\left(\frac{8}{6}\right)$ with East, so it makes $90° + 53°$ with North. It is represented by rt 143 fd 10.

Type in fd 8 rt 143 fd 10. This represents **F** + **G**. Is the result 6 units due East of the starting point? Check by returning to the starting point and typing in rt 90 fd 6.

Check some of the other results of this chapter.

14 Statistical measures

Averages

If you are given a set of numbers, it is useful to have a single number which tells you roughly how large the numbers are. This number is an **average**. It is also sometimes called a **measure of central tendency**, because it shows where the centre of the data is.

There are three commonly used averages. They are the **mean**, the **median** and the **mode**.

To find the mean of n numbers, add them and divide by n.

To find the median of n numbers, arrange them in order. If n is odd, take the middle number. If n is even, take the mean of the two middle numbers.

The mode of a set of numbers is the most common one.

> There may be more than one mode. If all the numbers are different, then there is no mode at all.

Example   The mean height of 20 men is 178 cm, and the mean height of 25 women is 160 cm. Find the mean height for all the people.

The *total* height for the men is 20×178 cm, which is 3560 cm. The total height for the women is 25×160 cm, which is 4000 cm. Add the totals, and divide by 45.

$$(3560 + 4000) \div 45 = 168$$

The mean height is 168 cm.

Exercise 14.1

1 Find the mean, median and mode (where it exists) of these sets of numbers.
  a 5, 4, 2, 6, 3, 5, 7, 10
  b 102, 112, 104, 106, 107, 108, 111, 106, 105
  c 0.04, 0.1, 0.06, 0.07, 0.04, 0.05, 0.06, 0.04
  d $-1.3, -1.7, 0, 0.5, -1.3, -0.5, 0.3$

2 A class of 12 boys and 15 girls took an exam. The mean for the boys was 43, and the mean for the girls was 46. Find the mean for the class as a whole.

3 In a five-week period a car dealership sold a mean of 13 cars per week. For the next three weeks they sold a mean of 16 cars per week. What was the mean for the eight-week period?

4 Over 60 summer days the mean midday temperature was 28 °C. For the first 20 days the mean was 29 °C. What was the mean for the final 40 days?

5 The mean income of a group of people is £18 000. The 12 men in the group have a mean income of £19 000. What is the mean income of the eight women?

6 Forty people were asked how often they had been to the cinema in the past month. The mean number of times was 1.6. If the mean for the 16 men was 1.9375, find the mean for the 24 women.

7 The mean of the numbers 3, 5, 6, 8, 9, x is 7.5. Find x.

8 The mean of the numbers $-4, -3, 0, x, 7, 12$ is 5. Find x.

9 The median of 2, 4, 7, x, 12, 14 is 8.5. Find x.

10 The median of $-1, 0, 2, 6, x, 13, 17, 20, 20, 21$ is 11.5. Find x.

11 The annual incomes of a group of 10 people were as follows. Find the mean and the median of the figures. Which average gives a better picture of the group as a whole? Give a reason.

£15 060	£18 430	£19 320	£17 550	£330 200
£16 430	£16 300	£17 450	£18 480	£19 100

12 Every week Jason has a test. The results for eight weeks are given below. Find the mean and median marks. Which average gives a better picture of Jason's ability? Give a reason.

27   32   35   29   30   0   31   29

Averages from frequency tables

If there are a lot of data, then it is not practical to list all the values. Instead they are often given in a **frequency table**. A frequency table can be either grouped or ungrouped.

If the data can take only a few possible values, then each value can be listed. The result is an **ungrouped** table.

For example, if you construct a table showing how many sisters people have, there are only a few possible values. The result might be an ungrouped table, like the one below.

number of sisters	0	1	2	3	4
frequency	26	18	8	4	1

If the data can take many possible values, then it is impractical to list every one. The possible values are put in intervals called groups. The result is a **grouped** table.

For example, if you want to construct a table showing how many marks people got in an exam out of 100, then there are too many possible values to be listed. The result might be a grouped table, as below.

marks	0–	20–	40–	60–	80–100
frequency	14	22	56	55	28

Remember:

Here, 20– means all the marks between 20 and 39, inclusive.

In an ungrouped table, all the information is there. We could, for example, list all the results of the survey about how many sisters people have. We would start with 26 0s, then 18 1s and so on. But, with a grouped table, we do not have all the information. We know that 14 people got marks between 0 and 19, but we don't know what the marks were. Hence any calculation from a grouped table is only approximate.

Example   For the first table above, find the mean, median and mode number of sisters.

Mean

There are 26 people with 0 sisters, 18 people with 1 sister and so on. The total number of sisters is

$$26 \times 0 + 18 \times 1 + 8 \times 2 + 4 \times 3 + 1 \times 4 = 50$$

The total number of people asked is $26 + 18 + 8 + 4 + 1$, which is 57. Divide 50 by 57.

The mean number of sisters is 0.877.

Median

If we listed all the numbers in order, there would be 26 0s, followed by 18 1s and so on. The middle of the 57 people would be the 29th person. In which column does the 29th person lie?

Notice that $26 < 29$, so the 29th person has at least one sister, and that $26 + 18 > 29$, so the 29th person has at most one sister. Therefore the 29th person would be in the 1s.

The median number of sisters is 1.

Mode

No calculation is required to find the mode. The number with the largest frequency, that is, the most common number, is 0.

The mode is 0.

Exercise 14.2

1 The table below gives the number of times that 100 drivers took the driving test before they passed. Find the mean, median and mode number of times.

number of times	1	2	3	4	5	6
frequency	38	29	18	9	4	2

2 The table below gives the number of A level passes of a group of university students. Find the mean, median and mode number of passes.

number of A levels	2	3	4	5
frequency	15	48	37	12

3 Pauline keeps a record of how many e-mails she receives each morning. The results over 50 days are below. Find the mean, median and mode number of e-mails.

number of e-mails	0	1	2	3	4	5
frequency	14	9	8	10	7	2

4 The number of goals scored in 40 soccer matches are given below. Find the mean, median and mode number of goals.

number of goals	0	1	2	3	4	5	6
frequency	10	10	8	5	3	2	2

5 In a survey, 50 people were asked how often they had eaten out during the past week. The results are below. Find the mean, median and mode number of times.

number of times	0	1	2	3	4	5
frequency	20	9	10	5	4	2

6 The mean of the data presented below is 7.2. Find x.

value	1	3	6	x	13
frequency	5	7	12	9	7

7 The mean of the data presented below is 1.58. Find x.

value	0	1	2	3	4
frequency	12	13	x	6	5

Mean from a grouped table

If data is presented in a grouped table, then we cannot recover all the information. So we can only make an estimate of the mean. In the example on page 198 about exam marks, there were 14 people who got marks between 0 and 19. We don't know whether they were nearer the 0 end or nearer the 19 end. The best we can do is to assume that they were evenly spread in the interval, and so averaging about 9.5 marks each.

Example   Estimate the mean mark for the students of the example on page 198.

Assume that the 14 people in the first group are spread evenly about 9.5. Hence the total mark for this group is 14×9.5. Similarly, assume that the 22 people in the second group are spread evenly about 29.5, the half-way point for the second group, and so on. It helps to put the data into a table as shown.

marks	frequency	half-way point	frequency × half-way point
0–19	14	9.5	$14 \times 9.5 = 133$
20–39	22	29.5	$22 \times 29.5 = 649$
40–59	56	49.5	$56 \times 49.5 = 2772$
60–79	55	69.5	$55 \times 69.5 = 3822.5$
80–100	28	90	$28 \times 90 = 2520$

Add up the right-hand column. The total mark is estimated at

$$14 \times 9.5 + 22 \times 29.5 + 56 \times 49.5 + 55 \times 69.5 + 28 \times 90 = 9896.5$$

The total number of students is $14 + 22 + 56 + 55 + 28$, which is 175. Divide 9896.5 by 175.

The mean mark is approximately 56.6.

> This answer is only approximate. Don't give it too high a degree of accuracy.

Exercise 14.3

1 The table below gives the maximum temperatures over 60 summer days. Estimate the mean temperature.

temperature (°C)	15–	20–	25–	30–35
frequency	6	17	23	14

2 The table below gives the times that 60 runners took to complete a race. Estimate the mean time.

time (minutes)	10–	12–	14–	16–18
frequency	5	18	25	12

3 A group of motorists was asked to record how many litres of petrol they used in a week. The results are below. Estimate the mean amount.

litres of petrol	0–	20–	40–	60–	80–100
frequency	12	8	9	6	4

4 The prices of 50 books in a bookshop are given in the table below. Estimate the mean price.

price (£)	4–	6–	8–	10–	12–	14–16
frequency	6	12	11	8	9	4

5 Is your answer to question 4 likely to be an overestimate or an underestimate? Give reasons.

Measures of spread

Averages do not tell us everything that we want to know about a set of data. The averages tell us where the centre of the data is, but they do not tell us how widely spread the data is about its centre. Look at these two diagrams.

The data have the same averages, but the numbers are much more widely spread in the first diagram than in the second. A number which tells us how wide the data is spread is a **measure of spread**.

You have already met two measures of spread, the range and the interquartile range.

Remember chapter 7.

The **range** of a set of data is the difference between the largest and the smallest. The **interquartile range** of a set of data is the difference between the upper and lower quartiles, which are the values which cut off the top quarter and the bottom quarter of the data.

The range can be a very unreliable measure of spread. It only takes two of the values, and hence gives no indication of how the values in between are spread. If there is an exceptionally large or exceptionally small value then the range is misleading.

The interquartile range is a more reliable measure of spread. Because it removes the top quarter and the bottom quarter of the data, it removes any exceptional value.

Exercise 14.4

1 For each of these sets of data, find the range.
   a 23, 26, 28, 29, 34
   b 0, 0.4, 0.5, 0.7
   c 123, 137, 144, 149, 158, 165

2 The weights of ten pedigree dogs and ten mongrel dogs were found. The results are below. For each set, find the mean and the range. Comment on the difference.

   pedigree dogs (kg)   3, 5, 6, 10, 11, 15, 20, 25, 29, 36
   mongrels (kg)   10, 11, 13, 16, 18, 18, 19, 20, 22, 23

3 Two keen golfers each go round the same course five times. Their results are below. For each set of results, find the mean and the range. Comment on the difference.

   Jill   92, 98, 87, 104, 99
   Edmund   92, 90, 89, 91, 93

   In golf, low scores are good!

4 In each of two towns, the prices of six houses were found. The results, in £1000s, are below. For each set, find the mean and the range. Comment on the difference.

Barchester	93	160	66	189	204	266
Fogton	56	65	49	68	70	58

5 A soft drinks firm has two machines for filling bottles with juice. The amount, in millilitres, delivered by each machine was tested five times. The results are below. For each machine, find the mean and the range. Comment on the difference.

machine A	499	500	502	496	503
machine B	491	483	533	502	491

6 Data is given in the frequency table below. Find the mean and the range.

value	0	1	2	3	4
frequency	10	15	17	6	2

SUMMARY

■ The **mean**, **median** and **mode** of a set of data are commonly used **averages**, or **measures of central tendency**.

■ The mean and the median can be found from a **frequency table**. If the table is **grouped**, then the results are approximate.

■ **Range** and **interquartile range** are both **measures of spread**.

Exercise 14A

1 Every day for eight days, Marcus measures the amount of time he spends connected to the internet. The results, in minutes, are below. Find the mean of these times.

30	15	5	24	2	12	157	11

2 Find the median of the times in question 1. Which average gives a better picture of Marcus' typical daily use of the internet?

3 The mean weekly incomes of a group of 20 people is £250. The mean of a group of 30 people is £280. What is the mean weekly income for all 50 people?

4 The mean of the numbers 3, 5, 7, 10, x, 17, 22, 30 is 13.375. Find x.

5 The table below gives the numbers of books read per week by 30 people. Find the mean number.

number of books	0	1	2	3	4
frequency	8	11	5	3	3

6 For the data in question 5, find the median number of books.

7 For the data in question 5, write down the mode number of books.

8 The table below gives the times spent sleeping by 60 people. Estimate the mean time.

time (hours)	6–	7–	8–	9–10
frequency	9	18	21	12

9 Find the range of the numbers in question 1.

10 Find the range of the numbers -3, -2, 0, 3, 5, 9.

Exercise 14B

1 Every morning when Karen puts her contact lenses in, she counts the number of attempts she makes before they are securely in her eyes. Her results over 10 days are as below. Find the mean number of attempts.

2	3	4	2	2	6	7	8	2	3

2 For the data in question 1, find the median and mode number of attempts.

3 Over a period of five weeks, the mean number of hours per day of rainy weather was 4. For the ten Saturdays and Sundays of the period, the mean was 6. What was the mean for the weekdays?

4 The median of the numbers 1, 3, 5, 6, y, 10, 11, 12 is 7.5. Find y.

5 A group of 50 people who regularly play the National Lottery were asked how many prizes they had won the previous year. Their answers are below. Find the mean number of prizes.

number of prizes	0	1	2	3	4	5
frequency	18	16	6	5	4	1

6 For the data in question 5, find the median number of prizes.

7 For the data in question 5, write down the mode number of prizes.

8 The minimum temperature was measured for the 59 days in January and February. The results are below. Estimate the mean minimum temperature.

temperature (°C)	−10 to −5	−5 to 0	0 to 5	5 to 10
frequency	6	17	24	12

9 Find the range of the numbers in question 1.

10 A data set contains −1 repeated 10 times, 0 repeated 5 times, and 1 repeated 10 times. Find the mean and the range of the data set.

Exercise 14C Ma1

You found the quartiles of large data sets in chapter 7, by using cumulative frequency graphs. One way of finding the quartiles for small data sets is the same as for finding the median. The median splits the data into two halves. The median of the lower half is the lower quartile, and the median of the upper half is the upper quartile. Take the numbers below.

   1   4   6   8   9   12   15   20

There are eight numbers. The median is 8.5. The lower half consists of 1, 4, 6 and 8, so the lower quartile is 5. Similarly the upper quartile is 13.5.

1 Find the quartiles for the following data sets.

   a   3   5   7   10   11   11   15   19
   b   12   13   17   20   22   23   30   32   33   38   40   42
   c   0   1   1   3   7   9   10   15   18
   d   20   22   27   28   29   32   35   40   41   42   48

2 Look at question 2 of exercise 14.4. Find the interquartile ranges for the two sets of data. Comment.

Exercise 14D Ma1

Find numerical data about a sport that you are interested in. Analyse the data using some of the methods of this chapter. For example, you could compare the consistency of two golfers or two runners by comparing the ranges of their performances. This could be done using data from the internet.

Exercise 14E Ma1

Statistical work involves a lot of number crunching. Computers can do most of the routine calculation. In particular, a spreadsheet package will contain functions to find the mean of a set of data. The notation used may differ between spreadsheets.

   Enter the numbers 5, 8, 9, 9, 10 and 19 in cells A1 to A6. In cell B1 enter @AVERAGE(A1..A6). The mean 10 should appear.

   Check some of your answers to the exercises in this chapter.

   There are many other functions available, which can be found by typing 'statistical function' in the index of the 'help' menu. Investigate these functions.

15 Transformations

Enlargements

An **enlargement** is a **transformation** which alters the size but not the shape of figures. The ratio by which the lengths are changed is the **scale factor**.

   In an enlargement, one point remains fixed. This is the **centre of enlargement**. We can do an enlargement by multiplying all distances from the centre by the scale factor.

Example Enlarge the triangle ABC by scale factor 2 from the point X.

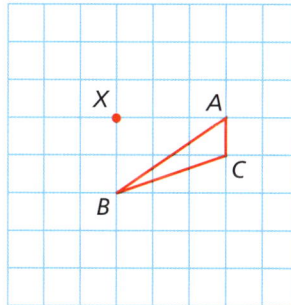

For each of the vertices, find the distance from X, then double this distance.

   A is 3 units to the right of X. So A' is 6 units to the right of X.
   B is 2 units below X. So B' is 4 units below X.
   C is 3 units to the right of X and 1 unit below. So C' is 6 units to the right of X and 2 units below.

The enlarged triangle is shown below.

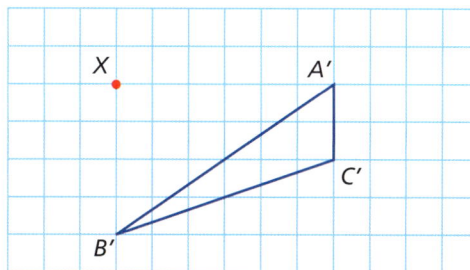

The scale factor of an enlargement can be less than 1. We still call it an enlargement, even though it makes figures smaller.

Example   Enlarge the quadrilateral *PQRS* from point $X(2, 2)$ by a scale factor of $\frac{1}{3}$.

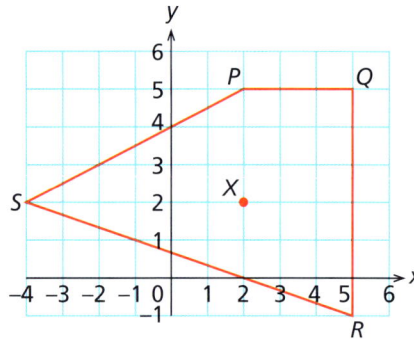

Here we set out the working in a table.

point	position relative to *X*	position relative to *X* after enlargement by $\frac{1}{3}$	new point
P(2, 5)	3 up	1 up	(2, 3)
Q(5, 5)	3 up, 3 to right	1 up, 1 to right	(3, 3)
R(5, −1)	3 down, 3 to right	1 down, 1 to right	(3, 1)
S(−4, 2)	6 to left	2 to left	(0, 2)

The enlarged quadrilateral is shown.

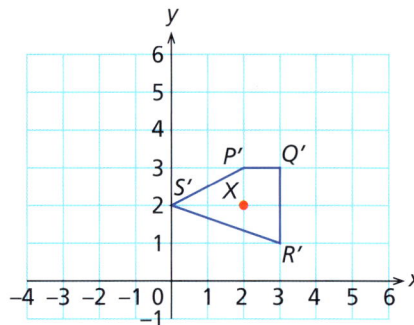

Exercise 15.1

1 Enlarge the triangle shown by a scale factor of 2 from the point (1, 2).

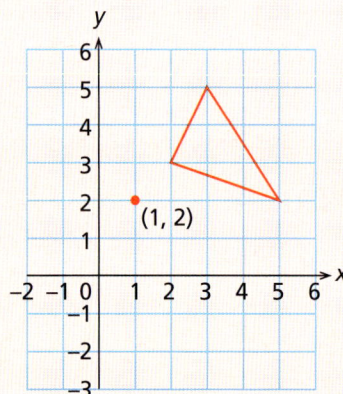

2 Enlarge the quadrilateral shown by a scale factor of 1.5 from the point (3, 1).

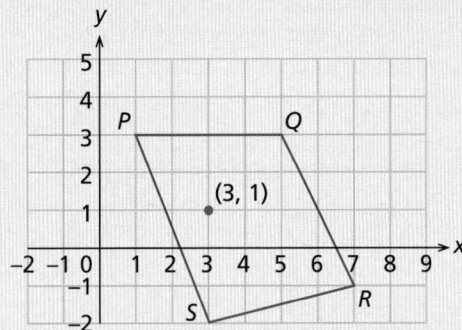

3 Enlarge the triangle shown by a scale factor of $\frac{1}{2}$ from the point (4, 3).

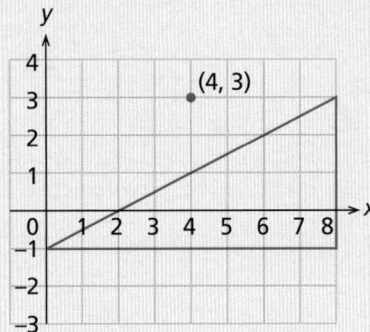

4 Enlarge the quadrilateral shown by a scale factor of $\frac{2}{3}$ from the point (3, 1).

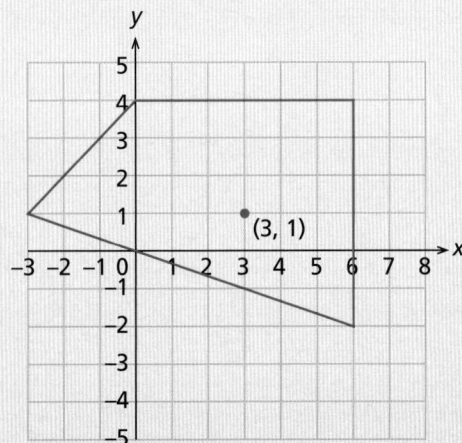

5 Plot the triangle with vertices at (1, 2), (3, 1) and (3, 3). Enlarge the triangle by a scale factor of 2 from (1, 4). Write down the coordinates of the new vertices.

6 Plot the triangle with vertices at (−2, 1), (0, 5) and (2, −1). Enlarge the triangle by a scale factor of 2.5 from (0, −1). Write down the coordinates of the new vertices.

7 Plot the quadrilateral with vertices at (2, −1), (−1, −1), (−1, 5) and (5, 5). Enlarge the triangle by a scale factor of $\frac{2}{3}$ from (2, 2). Write down the coordinates of the new vertices.

Negative scale factor

So far the scale factor has always been positive. If an enlargement has negative scale factor, then the distance from the centre is reversed by the enlargement. If a point is above the centre, then after enlargement it is below the centre. If a point is to the right of the centre, then after enlargement it is to the left.

Example Enlarge the triangle ABC from $X(4, 2)$ by a scale factor of -2.

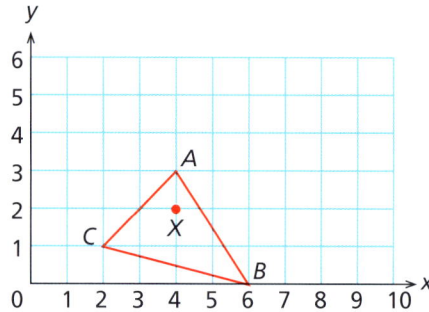

Fill in a table, making sure that the positions relative to X are reversed. So swap round *above* and *below*, and *right* and *left*.

point	position relative to X	position relative to X after enlargement by -2	new point
$A(4, 3)$	1 above	2 below	$(4, 0)$
$B(6, 0)$	2 right, 2 below	4 left, 4 above	$(0, 6)$
$C(2, 1)$	2 left, 1 below	4 right, 2 above	$(8, 4)$

The enlarged triangle is shown.

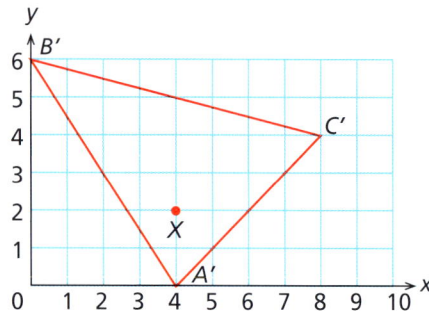

Exercise 15.2

1 Make a copy of the diagram, and enlarge the triangle shown by a scale factor of −2 from (3, 2).

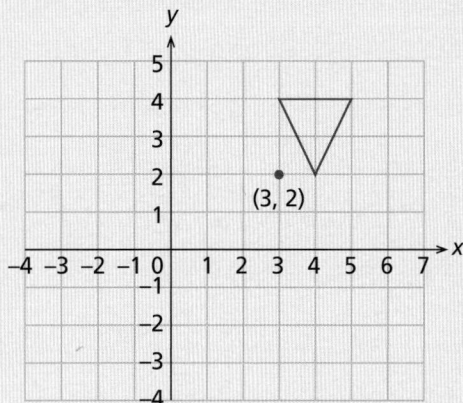

2 Make a copy of the shape on the diagram, and enlarge the triangle shown by a scale factor of −3 from (1, 1).

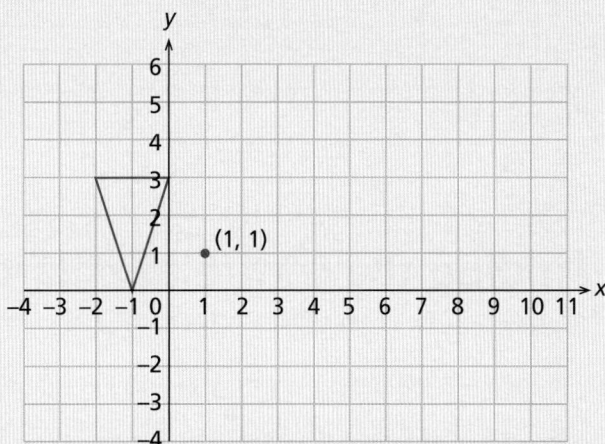

3 Make a copy of the diagram, and enlarge the triangle shown by a scale factor of −½ from (3, −1).

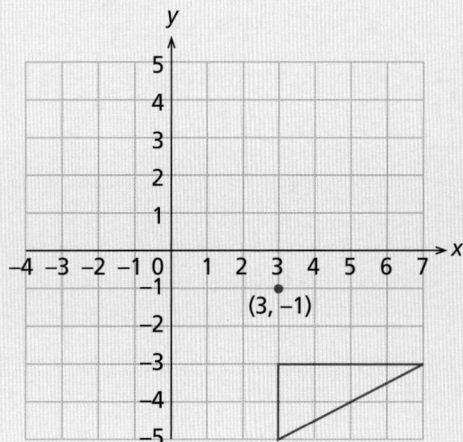

4 Make a copy of the diagram, and enlarge the triangle shown by a scale factor of $-\frac{2}{3}$ from $(3, 2)$.

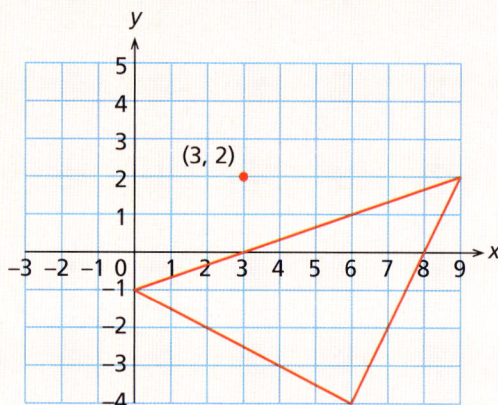

5 Plot the triangle with vertices at $(-2, 1)$, $(1, 4)$ and $(3, -1)$. Enlarge the triangle by a scale factor of -2 from $(0, -1)$. Write down the coordinates of the new vertices.

6 Plot the triangle with vertices at $(4, -1)$, $(2, 5)$ and $(-2, 1)$. Enlarge the triangle by a scale factor of $-\frac{1}{2}$ from $(-2, 3)$. Write down the coordinates of the new vertices.

7 Plot the quadrilateral with vertices at $(2, -1)$, $(4, 1)$, $(0, 5)$ and $(-2, -1)$. Enlarge the quadrilateral by a scale factor of -1.5 from $(4, 3)$. Write down the coordinates of the new vertices.

8 Plot ABC at $(1, 2)$, $(3, 5)$ and $(5, 1)$. Enlarge the triangle from $(0, 0)$ by a scale factor of -1. What other transformation is this equivalent to?

Finding the centre of enlargement

Suppose you are given a figure before and after enlargement. You can find the centre of enlargement X as follows.

Suppose A has been taken to A', and B has been taken to B'. Then X, A and A' all lie on a straight line. Similarly X, B and B' all lie on a straight line. So join A to A' and B to B'. Where AA' and BB' meet is X, the centre of enlargement.

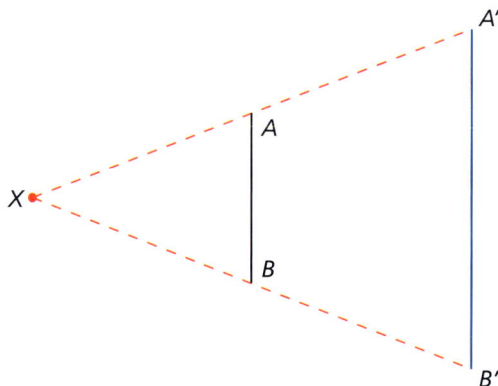

Example   In this diagram, triangle ABC has been enlarged to triangle $A'B'C'$. Find the centre of enlargement.

What is the scale factor of the enlargement?

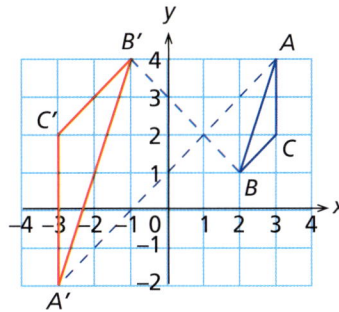

Join AA' and BB'. Notice that they cross at $(1, 2)$. As a check, notice that CC' also goes through $(1, 2)$.

The centre of enlargement is $(1, 2)$.

Notice that XA' is twice as long as XA, and that A' is on the other side of X from A.

The scale factor is -2.

Exercise 15.3

You will need:
• tracing paper

1 In this diagram one triangle is an enlargement of the other. Trace the diagram and find the centre of enlargement. What is the scale factor?

2 In this diagram one quadrilateral is an enlargement of the other. Trace the diagram and find the centre of enlargement. What is the scale factor?

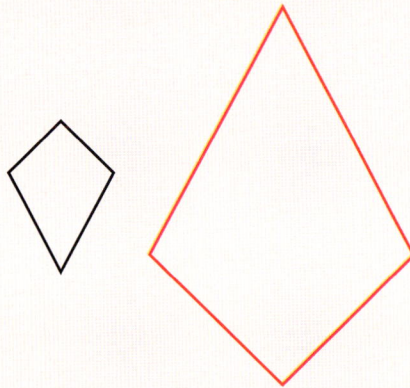

3 Plot ABC at $(1, 3)$, $(1, 1)$ and $(2, 3)$. This is enlarged to $A'B'C'$, at $(1, 2)$, $(1, -2)$ and $(3, 2)$. Plot this second triangle, and find the centre and scale factor of the enlargement.
4 The triangle with vertices at $(0, 4)$, $(6, 0)$ and $(-2, 2)$ is enlarged to the triangle with vertices at $(1, 3)$, $(4, 1)$ and $(0, 2)$. Plot the triangles, and find the centre and scale factor of the enlargement.
5 The triangle with vertices at $(1, 2)$, $(-1, -2)$ and $(3, -2)$ is enlarged to the triangle with vertices at $(2, 3)$, $(-1, -3)$ and $(5, -3)$. Plot the triangles, and find the centre and scale factor of the enlargement.
6 The quadrilateral with vertices at $(4, 4)$, $(7, 1)$, $(7, -2)$ and $(1, -2)$ is enlarged to the quadrilateral with vertices at $(-3, -3)$, $(-7, 1)$, $(-7, 5)$ and $(1, 5)$. Plot the quadrilaterals, and find the centre and scale factor of the enlargement.

Combinations of transformations

When you do one transformation after another, the result is a **combined transformation** (or **composition**). In many cases the combined transformation can be described as a single transformation. One simple case is this: if you do one **translation** after another, the result is a third translation, found by adding the **vectors** of the two original translations.

The translation $\begin{pmatrix} 3 \\ 4 \end{pmatrix}$ followed by the translation $\begin{pmatrix} -2 \\ 3 \end{pmatrix}$ is the translation $\begin{pmatrix} 1 \\ 7 \end{pmatrix}$.

An enlargement followed by another enlargement is normally an enlargement.

Example The triangle ABC is enlarged by a factor 2 from $(1, 4)$, then enlarged by a factor 1.5 from $(1, 2)$. Find the single equivalent enlargement.

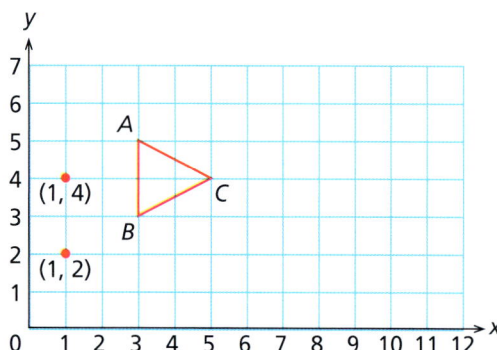

The effect of the two enlargements is shown. Notice that the sides of the final triangle are 3 times the sides of *ABC*. So the scale factor is 3. By joining up corresponding sides, we find that the centre of enlargement is at (1, 3.5).

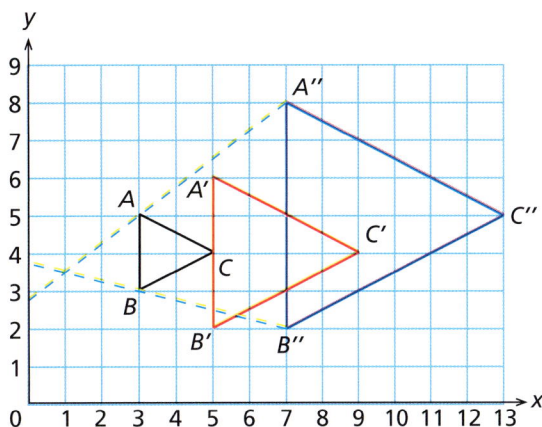

The combined transformation is an enlargement of scale factor 3 from (1, 3.5).

Note. For the scale factor, we do not need to do any drawing. Just multiply together the two original scale factors: $2 \times 1.5 = 3$.

Exercise 15.4

1 A translation with vector $\begin{pmatrix} 4 \\ -2 \end{pmatrix}$ is combined with a translation with vector $\begin{pmatrix} 3 \\ 7 \end{pmatrix}$. Write down the vector of the combined translation.

2 A translation with vector $\begin{pmatrix} -1 \\ 7 \end{pmatrix}$ is combined with a translation with vector $\begin{pmatrix} 2 \\ -3 \end{pmatrix}$. Write down the vector of the combined translation.

3 Suppose that the translations of questions 1 and 2 are done in the opposite order. Does this affect the final result?

4 Plot *ABC* at (1, 1), (2, 1) and (1, 2). Enlarge the triangle from centre (1, 1) by factor 2, then enlarge from (3, 3) with scale factor 3. Find the single equivalent enlargement.

5 Plot *PQR* at (2, 3), (4, 2) and (1, 1). Enlarge the triangle from centre (1, 3) by factor 2, then enlarge from (1, 5) with scale factor 1.5. Find the single equivalent enlargement.

> In these questions you are asked for a single transformation. Make sure you give only one transformation.

6 Plot *LMN* at (1, 1), (4, 3) and (5, 1). Enlarge the triangle from centre (2, 2) by factor −2, then enlarge from (0, 0) with scale factor −1. Find the single equivalent enlargement.

7 Plot *XYZ* at (4, 1), (−2, 3) and (−4, −1). Enlarge the triangle from centre (1, −1) by factor $\frac{1}{2}$, then enlarge from (0, 0) with scale factor 3. Find the single equivalent enlargement.

8 Combine the enlargements of question 4 in the opposite order, that is, do the enlargement from (3, 3) first, then the enlargement from (1, 1). Does this affect the final answer?

9 This example shows that the combination of two enlargements need not be another enlargement. Plot *ABC* at (1, 3), (3, 3) and (3, 1). Enlarge the triangle from (1, 1) by factor 2, then enlarge from (5, 5) by factor $\frac{1}{2}$. What single transformation is the combination equivalent to?

10 Give a rule for when the combination of two enlargements is not another enlargement.

Two reflections

Recall that a **reflection** is done by taking each point to the other side of a mirror line. With two reflections, the result depends on whether the mirror lines are parallel or not.

Example   The diagram shows a red triangle. Find its image after

a reflection in $x = 1$ then in $x = 3$
b reflection in $x = 0$ then in $y = x$.

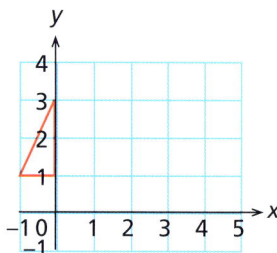

a The original red triangle is reflected in $x = 1$ to the blue triangle. This is then reflected in $x = 3$ to the green triangle. Notice that the final green triangle is 4 units to the right of the original red triangle. This is a translation.
   The combined transformation is a translation 4 units to the right.

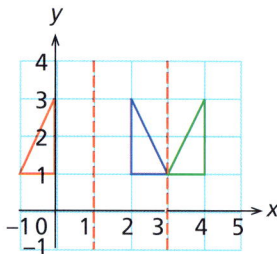

b The original red triangle is reflected in $x = 0$ to the blue triangle. This is then reflected in $y = x$ to the green triangle. Notice that the original red triangle has been rotated. The angle is 90° clockwise. The centre of rotation must be the intersection of the two reflection mirror lines.

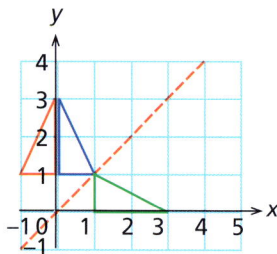

The combined transformation is a rotation, 90° clockwise, about $(0, 0)$.

Exercise 15.5

1 Plot a line segment AB at $(2, 3)$ and $(3, 3)$. Reflect first in $y = 4$, then in $y = 1$. What is the single transformation equivalent to the combination of reflections?

2 Plot a triangle at $(1, 1)$, $(2, 4)$ and $(3, 2)$. Reflect first in $x = 0$, then in $x = 1$. What is the single transformation equivalent to the combination of reflections?

3 Plot a triangle at $(2, 1)$, $(3, 5)$ and $(6, 1)$. Reflect first in $x = 0$, then in $y = 0$. What is the single transformation equivalent to the combination of reflections?

4 Plot a triangle at $(0, 3)$, $(-2, 5)$ and $(-1, 3)$. Reflect first in $x = 0$, then in $y = x$. What is the single transformation equivalent to the combination of reflections?

5 Plot a triangle at $(4, 1)$, $(3, 2)$ and $(6, 1)$. Reflect first in $y = 0$, then in $y = -x$. What is the single transformation equivalent to the combination of reflections?

6 Repeat question 1, but with the reflections in the opposite order, that is, reflect first in $y = 1$ then in $y = 4$. Is the result different?

7 Repeat question 4, but with the reflections in the opposite order. Is the result different?

Two rotations

Recall that a **rotation** is done by turning shapes through a certain angle about a fixed point, called the centre of rotation. A positive angle gives an anti-clockwise rotation, and a negative angle a clockwise rotation. If you are doing rotations on plain paper, it often helps to use tracing paper.

When two rotations are combined, the result depends on whether the two angles add up to $0°$ (or $360°$) or not.

Example Find the combined transformation equivalent to

a rotating through $90°$ anti-clockwise about $(0, 0)$, then $90°$ clockwise about $(4, 0)$

b rotating through $90°$ clockwise about $(0, 0)$, then $90°$ clockwise about $(0, 0)$.

The operations of part **a** are similar to 'walking' a heavy piece of furniture, when we move it along a room by successive rotations about one of its legs.

a Test on the red triangle as shown. It moves first to the blue triangle, then to the green triangle. The red triangle has moved 4 up and 4 to the right.

The combined transformation is a translation, with vector $\begin{pmatrix} 4 \\ 4 \end{pmatrix}$.

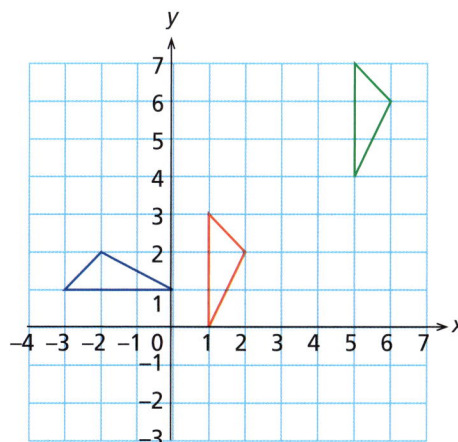

b In this case the triangle has been rotated, through $90° + 90°$, that is, through $180°$. The centre of rotation is still $(0, 0)$.

   The combined transformation is a rotation, of $180°$ about $(0, 0)$.

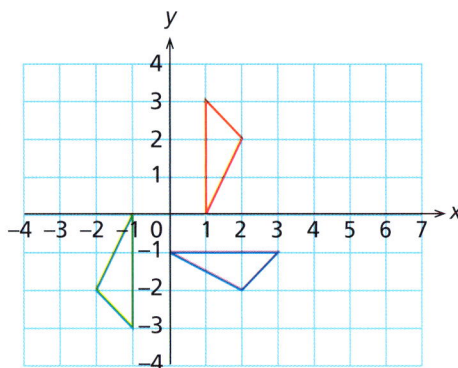

Exercise 15.6

1 Plot a triangle at $(0, 0)$, $(1, 1)$ and $(0, 3)$. Rotate it through $90°$ clockwise about $(0, 0)$, then through $90°$ anti-clockwise about $(1, -1)$. What single transformation is the combination equal to?

2 Plot a rectangle at $(0, 0)$, $(0, 2)$, $(4, 2)$, $(4, 0)$. Rotate the rectangle through $90°$ anti-clockwise about $(0, 2)$, then through $90°$ clockwise about $(0, 6)$. What single transformation is the combination equal to?

3 Rotate the triangle of question 1 through $90°$ about $(0, 0)$, then through a further $90°$ about $(0, 0)$. What single transformation is the combination equal to?

> **Remember:**
> *If no direction is specified, a rotation through a positive angle is anti-clockwise.*

4 Rotate the rectangle of question 2 through $90°$ about $(0, 2)$, then through a further $90°$ about $(0, 6)$. What single transformation is the combination equal to?

5 Repeat question 1, but with the rotations in the opposite order. Is the result the same as for question 1?

6 Repeat question 4, but with the rotations in the opposite order. Is the result the same as for question 4?

We can summarise the results that you should have discovered in the last three exercises.

- The combination of two translations is another translation.
- If the product of the scale factors of two enlargements is 1, then their combination is a translation.
- If the product of the scale factors of two enlargements is not 1, then their combination is an enlargement.
- If the mirror lines are parallel, the combination of two reflections is a translation.
- If the mirror lines are not parallel, the combination of two reflections is a rotation.
- If the angles add up to $0°$, the combination of two rotations is a translation.
- If the angles do not add up to $0°$, the combination of two rotations is a rotation.

Here are some miscellaneous problems, mixing different types of transformations.

Exercise 15.7

1 Plot a triangle at (1, 1), (3, 1) and (1, 4). Enlarge the triangle by factor 2 from (1, 1), then translate 2 to the right. What is the single transformation equivalent to this combination?
2 Repeat question 1, but with the transformations in the opposite order. Is the result the same?
3 Plot a triangle at (0, 2), (−1, 3) and (0, −3). Reflect the triangle in the line $x = 2$, then translate the result 3 units upwards. Write down the coordinates of the new vertices.
4 Plot a triangle at (1, 1), (3, 0) and (5, 2). Translate the triangle by the vector $\begin{pmatrix} 3 \\ -2 \end{pmatrix}$, then reflect the result in the line $y = 1$. Write down the coordinates of the new vertices.
5 Plot a rectangle at (0, 1), (0, 2), (2, 2), (2, 1). Rotate the rectangle through 90° about (0, 1), then translate 3 units to the right. What is the single transformation equivalent to this combination?
6 Repeat question 4, but with the transformations in the opposite order. Is the result the same?

7 A translation has vector $\begin{pmatrix} 4 \\ 0 \end{pmatrix}$. Find two reflections whose composition is equivalent to the translation.

8 Find two reflections whose composition is equivalent to a rotation of 180° about (1, 1).

Inverses of transformations

The **inverse** of a function is one which reverses its effect. So the inverse of the function x^3 is $\sqrt[3]{x}$, and the inverse of the function $2x + 1$ is $\frac{1}{2}(x - 1)$. Similarly, the inverse of a transformation reverses its effect. After combining a function with its inverse, every point is back where it started from.

Example   Find the inverse of the translation with vector $\begin{pmatrix} 4 \\ -3 \end{pmatrix}$.

This translation moves points 4 to the right and 3 down. The inverse does the opposite – it moves points 4 to the left and 3 up.

The inverse translation has vector $\begin{pmatrix} -4 \\ 3 \end{pmatrix}$.

Exercise 15.8

1 Find the vector of the translation which is inverse to the translation with vector $\begin{pmatrix} -3 \\ 5 \end{pmatrix}$.

2 Find the vector of the translation which is inverse to the translation with vector $\begin{pmatrix} 2 \\ -7 \end{pmatrix}$.

3 An enlargement has scale factor 2 and centre (1, 3). Apply the enlargement to the triangle with vertices at (1, 3), (1, 5) and (4, 3). Write down the inverse transformation.
4 An enlargement has scale factor −3 and centre (0, 0). Apply the enlargement to the triangle with vertices at (1, 2), (1, 4) and (5, 1). Write down the inverse transformation.
5 A reflection has mirror line $x = 2$. Apply the reflection to the line segment with endpoints (0, 4) and (1, −1). Write down the inverse transformation.
6 A reflection has mirror line $y = x$. Apply the reflection to the line segment with endpoints (3, 2) and (2, 0). Write down the inverse transformation.

7 A rotation is of angle 90° clockwise, about $(1, 4)$. Apply the rotation to the triangle with vertices at $(1, 4)$, $(3, 6)$ and $(3, 2)$. Write down the inverse transformation.

8 A rotation is of angle 90°, about $(1, 4)$. Apply the rotation to the line segment with ends at $(0, 2)$ and $(3, 0)$. Write down the inverse transformation.

9 You may have found that you can write down the inverses of these transformations without having to apply them to a particular shape. Write down the inverses of

a an enlargement of factor 3 from $(2, 5)$

b a reflection in the line $y = 3x$

c a rotation of 30° about $(2, -3)$.

Here is a summary of the results you should have discovered in the last exercise.

- The inverse of a translation is the translation in the opposite direction. Multiply its vector by -1.

- The inverse of an enlargement with scale factor k about C is an enlargement, scale factor $\dfrac{1}{k}$, also about C.

- The inverse of a reflection is the same reflection.
- The inverse of a rotation of $\theta°$ about C is a rotation of $-\theta°$ about C.

Finding the centre of rotation

In the diagram, the line segment AB has been rotated to $A'B'$. To find the angle of rotation we measure angles; how do we find the centre of rotation C?

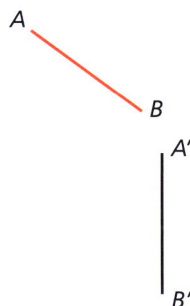

During rotation, distances from the centre remain constant. So $AC = A'C$, and $BC = B'C$. The centre C is equidistant from A and A', and from B and B'. So C will lie on the perpendicular bisectors of AA' and BB'. The diagram shows where C is.

Remember chapter 5.

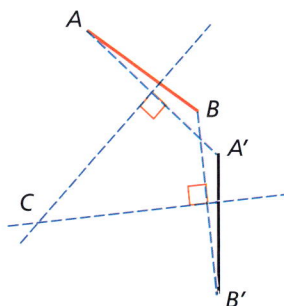

Example In a rotation, ABC is taken to $A'B'C'$. Find the centre of rotation.

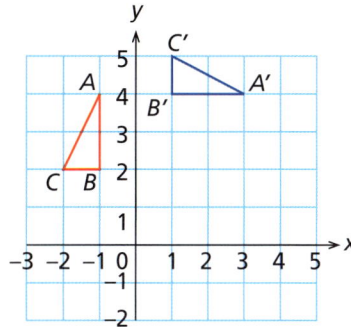

The line AA' is horizontal, with centre at $(1, 4)$. Hence the perpendicular bisector of AA' is the vertical line through $(1, 4)$, which is $x = 1$ (red in the diagram). Similarly the perpendicular bisector of BB' goes through $(0, 3)$, with gradient -1 (blue in the diagram).

These lines meet at $(1, 2)$.

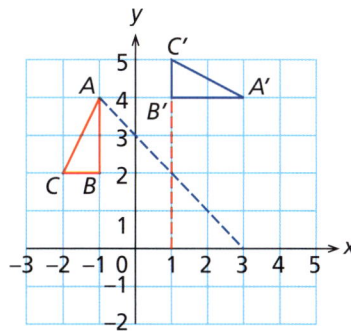

The centre of rotation is $(1, 2)$.

Exercise 15.9

You will need:
• tracing paper

1 In this diagram triangle T has been rotated to triangle S. Trace the diagram, and find the centre and angle of the rotation.

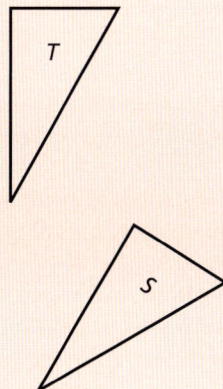

2 In this diagram rectangle X has been rotated to rectangle Y. Trace the diagram, and find the centre and angle of the rotation.

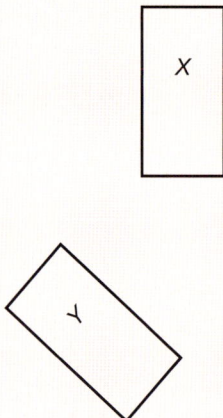

3 The triangle with vertices at $(2, 2)$, $(2, 4)$ and $(1, 4)$ is rotated to the triangle with vertices at $(1, 3)$, $(-1, 3)$ and $(-1, 2)$. Plot the triangles, and find the centre and angle of the rotation.

4 The line segment between $(2, 3)$ and $(4, 3)$ is rotated to the segment between $(2, -1)$ and $(2, -3)$. Plot the segments, and find the centre and angle of the rotation.

5 The line segment between $(1, 6)$ and $(6, 1)$ is rotated to the segment between $(5, 4)$ and $(4, -3)$. Plot the segments, and find the centre and angle of the rotation.

6 The line segment between $(4, -10)$ and $(-9, 3)$ is rotated to the segment between $(-1, -9)$ and $(-8, 8)$. Plot the segments, and find the centre and angle of the rotation.

SUMMARY

■ An **enlargement** alters the sizes of figures without altering their shapes. The **centre of enlargement** is the point which remains fixed. Distances from the centre are multiplied by a number, the **scale factor** of the enlargement.

■ If the scale factor is negative, then the distances to points from the centre of enlargement are reversed.

■ To find the centre of enlargement, join up corresponding points (from the original shape and its enlargement) and find the intersection.

■ A **combined transformation** (or **composition**) is done by taking one transformation after another.

■ The **inverse** of a transformation returns points to their original positions.

■ To find a centre of rotation, find the intersection of the perpendicular bisectors of two pairs of corresponding points.

Exercise 15A

1 Plot points at $(0, -1)$, $(0, 5)$ and $(6, 3)$. Enlarge the triangle formed by a scale factor of $\frac{1}{2}$ with centre $(2, 1)$.

2 Plot points at $(1, 5)$, $(2, 3)$ and $(6, 6)$. Enlarge the triangle formed by a scale factor of -2 with centre $(1, 1)$.

3 The diagram shows two shapes which are enlargements of each other. Trace the diagram and find the centre of enlargement.

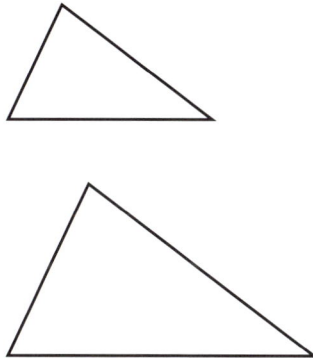

4 Write down the vector of the translation which is the combination of the translations with vectors $\begin{pmatrix} 2 \\ -1 \end{pmatrix}$ and $\begin{pmatrix} -5 \\ -3 \end{pmatrix}$.

5 Let $\triangle ABC$ have vertices at $(1, 3)$, $(4, 6)$ and $(4, 0)$. Find the image of this triangle after it has been translated by the vector $\begin{pmatrix} -2 \\ 3 \end{pmatrix}$ and then enlarged from $(0, 0)$ by a scale factor of 2. Describe the single transformation equivalent to this combination.

6 Repeat question 5, reversing the order of the transformations.

7 The triangle of question 5 is reflected in the line $y = x$ and then reflected in the x-axis. Find the single transformation equivalent to this combination.

8 The triangle of question 5 is rotated 90° clockwise about $(4, 0)$, and then 90° anti-clockwise about $(0, 0)$. Find the single transformation equivalent to this combination.

9 Write down the inverse transformation to a reflection in the line $y = 17x - 2\frac{1}{3}$.

10 The triangle with vertices at $(0, 1)$, $(-2, 1)$ and $(-1, 2)$ is rotated to the triangle with vertices at $(2, 1)$, $(4, 1)$ and $(3, 0)$. Plot the triangles, and find the centre and angle of rotation.

Exercise 15B

1 Plot the rectangle with points $(-1, -3)$, $(-1, 1)$, $(2, 1)$ and $(2, -3)$. Enlarge the rectangle from $(1, 0)$ with scale factor $\frac{1}{2}$.

2 Enlarge the rectangle of question 1 from $(-1, 0)$ with scale factor -1.

3 The triangle with vertices $(1, 1)$, $(2, 2)$ and $(3, 1)$ is enlarged to the triangle with vertices $(1, -1)$, $(4, 2)$ and $(7, -1)$. Plot the triangles, and find the centre and scale factor of the enlargement.

4 Find the vector of the combination of the translations with vectors $\begin{pmatrix} 5 \\ 4 \end{pmatrix}$ and $\begin{pmatrix} -1 \\ 3 \end{pmatrix}$.

5 The triangle of question 3 is enlarged by factor 3 from the point $(1, 1)$, and then enlarged by factor $\frac{1}{3}$ from the point $(4, 1)$. What single transformation is equivalent to the two enlargements?

6 The triangle of question 3 is reflected in the line $x = 4$, and then in the line $x = 1$. Find the single transformation equivalent to the combination.

7 Repeat question 6, reversing the order of the reflections.

8 A triangle has vertices at (2, 2), (2, 4) and (3, 4). It is rotated 90° about (2, 1), and then a further 90° about (1, 1). Find the single transformation equivalent to the two rotations.

9 Write down the inverse of a rotation of 38° about (4, 7).

10 In the diagram, the left-hand shape has been rotated to the right-hand shape. Trace the diagram, and find the centre and angle of rotation.

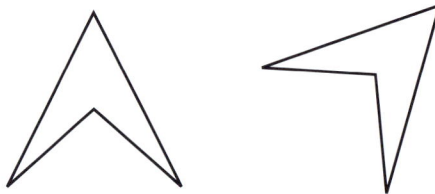

Exercise 15C Ma1

A wallpaper pattern consists of a motif repeated horizontally and vertically. The motif could be repeated by simple translation, or by reflection or rotation.

Suppose the motif is as shown. The diagrams show how it is repeated.

a By horizontal translation, but vertical reflection.

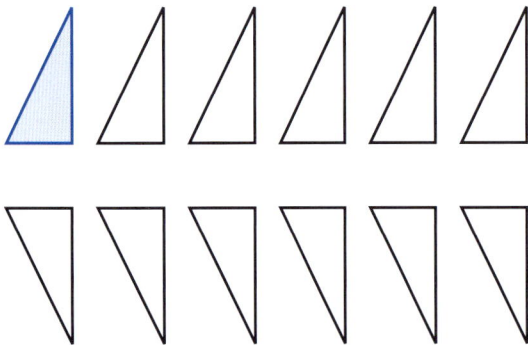

b By horizontal rotation and vertical translation.

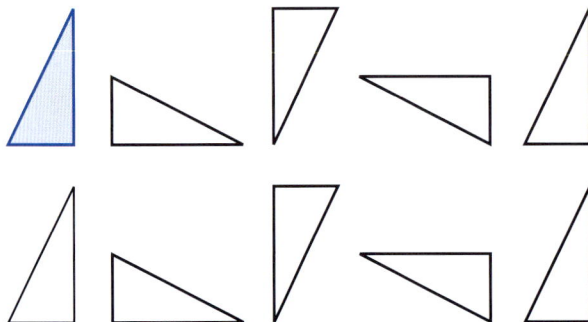

c By rotation on a triangular grid.

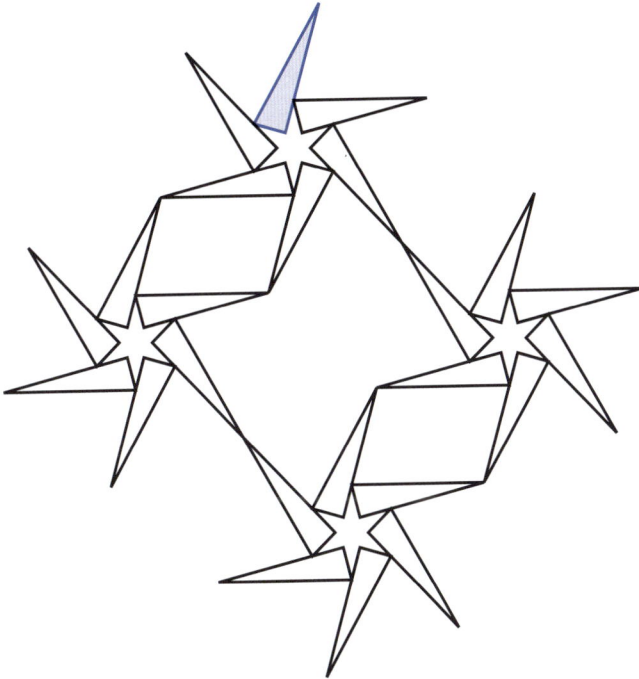

Find as many different patterns as you can. It has been shown that there are essentially 17 different patterns of wallpaper. How many did you get?

The tiles in the Alhambra, in Granada in Spain, have examples of all 17 types of pattern.

Exercise 15D Ma1

All the transformations in this chapter have been in two dimensions. It is possible to have transformations in three (or more) dimensions.

1 In two dimensions, we speak of reflection in a line. What is a three-dimensional reflection?
2 In two dimensions, we speak of rotation about a point. What is a three-dimensional rotation?
3 A three-dimensional reflection reverses left and right. (If you are right-handed, then your image in a mirror is a left-handed person.) Does a three-dimensional enlargement, scale factor -1, change a left hand to a right hand?
4 A mirror reverses left and right, and it also reverses back and front. Why doesn't it reverse top and bottom?

Exercise 15E Ma1

A geometry package or a drawing package can do all the transformations of this chapter. Details will vary between packages, but in all cases you should be able to:

You will need:
• computer
• geometry package such as *Cabri* or drawing package such as *Corel Draw*

1 select the object you want to transform
2 select the transformation (rotation, reflection, and so on)
3 select the details of the transformation (centre of rotation, slope of mirror line, and so on)
4 perform the transformation, keeping the original shape if you want to.

1 Look at the example about reflections of a triangle on page 214. Draw a triangle on the screen. Reflect it in the two pairs of lines as given. In each case check that the final result is as given.
   Check your answers to some of the other questions in the chapter.
2 You can use transformations to create patterns. In particular, a computer can do rotations about angles other than 90° or 180°, which are very hard to do by hand.
   Draw a shape on the screen. Rotate it about 10° about a fixed point. Repeat the rotation until you are back to where you started from. You should have a pattern with rotational symmetry.
   Draw some other patterns of your own.
3 Recall that a **tessellation** is a covering of the plane with equal shapes. Obviously rectangles tessellate, but what is less obvious is that *any* quadrilateral tessellates.
   Draw any quadrilateral on the screen. Show how to tessellate with transformations of this quadrilateral.

Hint: the sum of the angles of a quadrilateral is 360°. The sum of the angles round any point is also 360°.

16 Algebra

Changing the subject

You may have recognised these formulae. They are for converting between temperature in Celsius and in Fahrenheit. The first formula converts Fahrenheit to Celsius, and the second converts Celsius to Fahrenheit.

In a **formula** with an = sign, the **subject** of the formula is the letter which is expressed in terms of the other letters. In the formula $C = \frac{5}{9}(F - 32)$, for example, the subject of the formula is C.

When we rearrange a formula so that a different letter is the subject, we **change the subject** of the formula. The formula above can be rearranged to

$$F = \tfrac{9}{5}C + 32$$

Note that F is now the subject of the formula.

Changing the subject of a formula is very similar to solving an equation. You must be sure to do the same operations on both sides of the equals sign. You must also be careful to do the operations in the correct order.

Examples In the formula $y = 3x - 7$, change the subject to x.

In the formula, x has been multiplied by 3 and then 7 has been subtracted. To undo these operations, reverse them in the opposite order. So add 7 to both sides, then divide by 3.

$$y + 7 = 3x$$
$$\tfrac{1}{3}y + \tfrac{7}{3} = x$$

$x = \tfrac{1}{3}y + 2\tfrac{1}{3}$.

In the formula $p = \tfrac{1}{4}q + 8$, change the subject to q.

Subtract 8 from both sides, then multiply by 4.

$$p - 8 = \tfrac{1}{4}q$$
$$4p - 32 = q$$

$q = 4p - 32$.

In the formula $m = 6(n - 8)$, change the subject to n.

Because of the brackets, 8 has been subtracted from n first and then the result has been multiplied by 6. To undo these operations, we divide by 6 first and then add 8.

$$\tfrac{1}{6}m = n - 8$$
$$\tfrac{1}{6}m + 8 = n$$

$n = \tfrac{1}{6}m + 8$.

Note. In this example, we could expand the brackets and then proceed as in the previous examples.

$$m = 6n - 48$$
$$m + 48 = 6n$$
$$\tfrac{1}{6}m + 8 = n$$

This gives the same answer, but it takes one more step.

Exercise 16.1

In questions 1–17, change the subject of the formula to the letter in brackets.

1 $y = 2x + 5$ (x)   **2** $m = 5n - 3$ (n)   **3** $p = 7q + 3$ (q)

4 $a = 1 - 7b$ (b)   **5** $m = 3 - 2n$ (n)   **6** $y = \tfrac{1}{3}x + 2$ (x)

7 $r = \tfrac{1}{5}t - 4$ (t)   **8** $m = 0.1p - 2$ (p)   **9** $g = 0.4h + 6$ (h)

10 $y = 3(x + 4)$ (x)   **11** $m = 4(n - 3)$ (n)   **12** $c = \tfrac{1}{4}(d - 2)$ (d)

13 $p = 0.2(q + 7)$ (q)   **14** $f = 2(3 - 5g)$ (g)   **15** $a = 3(2 - 5b)$ (b)

16 $u = \tfrac{1}{2}(5 - 3v)$ (v)   **17** $y = 0.1(3 - 2x)$ (x)

18 To hire a van, there is a fixed payment of £20 and each day costs £25. If I hire the van for d days, the cost £C is given by $C = 25d + 20$. Make d the subject of this formula.

19 A gas bill consists of £15 standing charge plus £0.01 per unit used. If I use u units the bill is £B.
  a Find a formula giving B in terms of u.
  b Make u the subject of this formula.

20 A money changer changes £ to $ at a rate of $1.6 per £, and then takes a commission of $5. Suppose I get $$d$ for £p.
  a Find a formula giving d in terms of p.
  b Make p the subject of this formula.

21 Another money changer asks for a commission of £3 first, and then changes £ to $ at the same rate. Suppose I get $$d$ for £p.
  a Find a formula giving d in terms of p.
  b Make p the subject of this formula.

Other examples of changing the subject involve multiplying or dividing by letters.

Examples   Make x the subject of the formula $y = 4ax$.

Divide both sides by 4 and by a.

$$\frac{y}{4a} = x$$

$$x = \frac{y}{4a}.$$

Make x the subject of the formula $y = mx + c$.

Subtract c, then divide by m.

$$y - c = mx$$

$$x = \frac{y - c}{m}.$$

This is the equation of a straight line.

Make a the subject of the formula $b = \dfrac{3}{a + 2}$.

Multiply both sides by $(a + 2)$, then divide by b. Finally subtract 2.

$$b(a + 2) = 3$$

$$a + 2 = \frac{3}{b}$$

$$a = \frac{3}{b} - 2.$$

Exercise 16.2

In questions 1–19, change the subject to the letter in brackets.

1 $m = an + 3$ (n) **2** $y = kx - 2$ (x)

3 $y = 3zx$ (x) **4** $m = 5kn$ (n)

5 $y = \frac{1}{2}kx$ (x) **6** $m = 0.2an$ (n)

7 $u = kv + w$ (v) **8** $m = an + b$ (b)

9 $y = kx - a$ (x) **10** $p = rq - s$ (q)

11 $y = a(x + z)$ (x) **12** $m = b(n + l)$ (l)

13 $a = k(b - m)$ (b) **14** $r = s(t - v)$ (v)

15 $M = N(P - T)$ (T) **16** $y = \dfrac{2}{x - 3}$ (x)

17 $m = \dfrac{p}{n + 4}$ (n) **18** $f = \dfrac{6}{2 - g}$ (g)

19 $h = \dfrac{1}{1 - 2k}$ (k)

20 A formula giving velocity v in terms of initial velocity u, time t and acceleration a is $v = u + at$. Change the subject to

     a u **b** t.

21 The area of a triangle is given by the formula $A = \frac{1}{2}bh$. Change the subject of the formula to b.

22 The surface area of a cylinder is given by the formula $A = 2\pi r(r + h)$. Change the subject of the formula to h.

Other examples of changing the subject involve squaring or taking the square root. These are inverse functions, so to get rid of a square root sign you square, and vice versa. So if $y = x^2$, then $x = \sqrt{y}$. If $m = \sqrt{n}$, then $n = m^2$.

Examples   Make r the subject of the formula $A = \pi r^2$.

> This is the formula for the area of a circle.

Divide both sides by π. Then to change r^2 to r, take the square root of both sides.

$$\frac{A}{\pi} = r^2$$

$$\sqrt{\frac{A}{\pi}} = r$$

$$r = \sqrt{\frac{A}{\pi}}.$$

Make a the subject of the formula $c = \sqrt{a^2 + b^2}$.

To remove the square root sign, square both sides. Subtract b^2 from both sides, then take the square root.

$$c^2 = a^2 + b^2$$

$$c^2 - b^2 = a^2$$

$$\sqrt{c^2 - b^2} = a$$

$$a = \sqrt{c^2 - b^2}.$$

Exercise 16.3

In questions 1–28, change the subject to the letter in brackets.

1 $y = 6x^2$ (x) **2** $m = 4n^2$ (n) **3** $p = aq^2$ (q)

4 $k = \frac{1}{2}j^2$ (j) **5** $a = 0.1b^2$ (b) **6** $c = \frac{2}{3}d^2$ (d)

7 $y = \dfrac{x^2}{a}$ (x) **8** $m = \dfrac{2n^2}{b}$ (n) **9** $p = q^2 - 2$ (q)

10 $y = x^2 + a$ (x) **11** $m = n^2 - b$ (n) **12** $s = 3 - r^2$ (r)

13 $p = a - q^2$ (q) **14** $y = 2x^2 + a$ (x) **15** $m = 3n^2 - b$ (n)

16 $p = aq^2 + b$ (q) **17** $a = kb^2 - m$ (b) **18** $y = \sqrt{2x}$ (x)

19 $a = \sqrt{kb}$ (b) **20** $y = \sqrt{\dfrac{x}{a}}$ (x) **21** $m = \sqrt{\dfrac{n}{3}}$ (n)

22 $m = \sqrt{n + 3}$ (n) **23** $y = \sqrt{x - a}$ (x) **24** $a = \sqrt{3 - b}$ (b)

25 $y = \sqrt{x^2 - z^2}$ (x) **26** $m = \sqrt{n^2 + k^2}$ (n) **27** $v = \sqrt{u^2 - w^2}$ (w)

28 $a = \sqrt{b^2 - c^2}$ (c)

29 If an electrical circuit has current I and resistance R, then its power P is given by $P = I^2R$. Make I the subject of this formula.

30 The volume of a cylinder is given by the formula $V = \pi r^2 h$. Make r the subject of this formula.

31 If an acceleration a acts over a distance s, the velocity changes from u to v, where $v^2 = u^2 + 2as$. Make u the subject of this formula.

> This is like **collecting like terms** when simplifying an expression.

Formulae in which the new subject occurs more than once

In some formulae, the new subject occurs more than once. For these formulae, collect together the terms which include the subject, and factorise.

Examples   Make x the subject of the formula $ax + 3 = 7 - bx$.

Collect the x terms on the left and the numbers on the right. Then factorise and divide.

$$ax + bx = 7 - 3$$
$$x(a + b) = 4$$

$$x = \frac{4}{a + b}.$$

Make T the subject of the formula $k = \dfrac{T + 5}{T - 7}$.

Multiply both sides by $(T - 7)$, expand the brackets and then proceed as above.

$$k(T - 7) = T + 5$$
$$kT - 7k = T + 5$$
$$kT - T = 5 + 7k$$
$$T(k - 1) = 5 + 7k$$

> Note that we need 1 inside the brackets, as there is a T term on its own.

$$T = \frac{5 + 7k}{k - 1}.$$

Exercise 16.4

In these questions change the subject to the letter in brackets.

1 $ay + 1 = by + 2$ (y) **2** $km + 3 = lm - 7$ (m)
3 $ax + 1 = 3x - 2$ (x) **4** $5y - 6 = ky - 8$ (y)
5 $5m + n = km$ (m) **6** $ak = 2 - bk$ (k)
7 $ax + b = cx + d$ (x) **8** $ka + m = la - n$ (a)
9 $xp + 1 = yp + zp$ (p) **10** $ax + m = n + bx - cx$ (x)

11 $y = \dfrac{x+1}{x+2}$ (x) **12** $m = \dfrac{n-3}{n+2}$ (n)

13 $p = \dfrac{q+4}{q-7}$ (q) **14** $R = \dfrac{3S}{S-4}$ (S)

15 $a = \dfrac{b+c}{b}$ (b) **16** $m = \dfrac{n+a}{n-b}$ (n)

17 $f = \dfrac{3g+7}{2g-3}$ (g) **18** $w = \dfrac{z+i}{z-2i}$ (z)

Algebraic fractions

In an **algebraic fraction**, the numerator and denominator may involve letters as well as numbers. The rules for algebraic fractions are the same as for ordinary numerical fractions.

Simplification

You can **simplify** a fraction, whether numerical or algebraic, by multiplying or dividing top and bottom by the same amount.

Example   Simplify the fraction $\dfrac{3x^2}{6xy}$, where $x \neq 0$.

Both numerator and denominator have $3x$ as a factor.

$$3x^2 = 3x \times x \qquad 6xy = 3x \times 2y$$

Divide top and bottom by $3x$ (to cancel $3x$).

$$\frac{3x^2}{6xy} = \frac{3 \times x \times x}{6 \times x \times y} = \frac{x}{2y}$$

Exercise 16.5

Simplify these fractions. Assume that none of the unknowns is 0.

1 $\dfrac{6}{2x}$ **2** $\dfrac{3x}{2x}$

3 $\dfrac{4a^2}{6ab}$ **4** $\dfrac{abc}{bcd}$

5 $\dfrac{x^2 y}{xy^2}$

6 $\dfrac{6a^2 b^3}{9a^3 b^2}$

7 $\dfrac{5xyz}{10x^2 y}$

8 $\dfrac{15m^2 n}{25mn^2}$

9 $\dfrac{xy + x}{zy + z}$

10 $\dfrac{3x + 9}{2x + 6}$

(Hint: factorise top by x and bottom by z.)

11 $\dfrac{4x - 6y}{15y - 10x}$

12 $\dfrac{x^2 + 5x + 6}{x^2 + 6x + 8}$

(Hint: factorise top and bottom.)

13 $\dfrac{x^2 - 3x - 10}{x^2 + 2x - 35}$

14 $\dfrac{x^2 - 4}{x^2 - 4x + 4}$

Arithmetic of algebraic fractions

Algebraic fractions are multiplied or divided in the same way as numerical fractions.

Examples Write as a single fraction: $\dfrac{2x}{3y^2} \times \dfrac{5xy}{4z}$.

The product of the numerators is $10x^2 y$, and the product of the denominators is $12y^2 z$. Divide top and bottom by $2y$ to simplify.

$$\frac{2x}{3y^2} \times \frac{5xy}{4z} = \frac{10x^2 y}{12y^2 z} = \frac{5x^2}{6yz}$$

$$\frac{2x}{3y^2} \times \frac{5xy}{4z} = \frac{5x^2}{6yz}.$$

Remember:

To multiply fractions, multiply the numerators and multiply the denominators. To divide fractions, turn the second fraction upside down and multiply.

Note. You might find it easier to cancel before multiplying.

$$\frac{2x}{3y^2} \times \frac{5xy}{4z} = \frac{\cancel{2}x}{3y^{\cancel{2}}} \times \frac{5x\cancel{y}}{\cancel{4}z} = \frac{5x^2}{6yz}$$
$$2$$

Write as a single fraction: $\dfrac{x + y}{a} \div \dfrac{2x + 2y}{b}$.

Turn the second fraction upside down and multiply. Factorise and cancel.

$$\frac{x + y}{a} \div \frac{2x + 2y}{b} = \frac{x + y}{a} \times \frac{b}{2x + 2y}$$
$$= \frac{b(x + y)}{a(2x + 2y)}$$
$$= \frac{b(x + y)}{2a(x + y)}$$
$$= \frac{b}{2a}$$

$$\frac{x + y}{a} \div \frac{2x + 2y}{b} = \frac{b}{2a}.$$

Exercise 16.6

Write the following as single fractions, simplifying your answers as far as possible.

1 $\dfrac{2a}{3b} \times \dfrac{6a}{10b}$

2 $\dfrac{4x}{21y} \times \dfrac{7a}{6b}$

3 $\dfrac{5x}{2y} \times \dfrac{3y}{15x}$

4 $\dfrac{4m}{3n} \times \dfrac{9n}{6m}$

5 $\dfrac{2m}{3n} \times \dfrac{3m}{5n}$

6 $\dfrac{4x}{5y} \times \dfrac{10x}{3y}$

7 $\dfrac{4a^2}{3b^2} \times \dfrac{2b}{3a}$

8 $\dfrac{2x^2}{3xy} \times \dfrac{6y^2}{10x}$

9 $\dfrac{x^2+x}{y} \times \dfrac{1}{x+1}$

10 $\dfrac{a}{b} \div \dfrac{a}{c}$

11 $\dfrac{x}{3} \div \dfrac{x}{4}$

12 $\dfrac{m}{3x} \div \dfrac{m}{5x}$

13 $\dfrac{x^2}{3} \div \dfrac{xy}{2}$

14 $\dfrac{x^2}{yz} \div \dfrac{y^2}{xz}$

15 $\dfrac{3xy}{4z^2} \div \dfrac{15y^2}{28xz}$

16 $\dfrac{3}{x+1} \div \dfrac{2}{xy+y}$

17 $\dfrac{a}{x^2+3x+2} \div \dfrac{b}{x^2+5x+6}$

Addition and subtraction

Algebraic fractions are added or subtracted in the same way as numerical fractions. If the fractions have the same denominator, then add or subtract the numerators. If the fractions have different denominators, then adjust the fractions so they have the same denominator. Then add or subtract the numerators. To add $\frac{1}{2}$ and $\frac{1}{3}$, for example, put them over the common denominator of 6.

$$\tfrac{1}{2} + \tfrac{1}{3} = \tfrac{3}{6} + \tfrac{2}{6} = \tfrac{5}{6}$$

Example Write as a single fraction: $\dfrac{x}{2} + \dfrac{y}{3}$.

The least common multiple of 2 and 3 is 6. Convert the fractions, then add the numerators.

$$\frac{x}{2} + \frac{y}{3} = \frac{3x}{6} + \frac{2y}{6} = \frac{3x+2y}{6}$$

$$\frac{x}{2} + \frac{y}{3} = \frac{3x+2y}{6}.$$

Exercise 16.7

Write the following as single fractions, simplifying your answers as far as possible.

1 $\dfrac{x}{3} - \dfrac{y}{3}$

2 $\dfrac{a}{3} + \dfrac{b}{5}$

3 $\dfrac{x}{2} + \dfrac{x}{8}$

4 $\dfrac{y}{15} - \dfrac{y}{9}$

5 $\dfrac{m}{3} - \dfrac{n}{2}$

6 $\dfrac{a+1}{2} + \dfrac{a+2}{3}$

7 $\dfrac{x-4}{4} + \dfrac{x+3}{3}$

8 $\dfrac{y+7}{6} - \dfrac{y+3}{4}$

9 $\dfrac{2x+3}{5} + \dfrac{3x+2}{4}$

10 $\dfrac{a+7}{3} - \dfrac{a-3}{2}$

11 $\dfrac{x+2y}{4} - \dfrac{x-3y}{3}$

12 $\dfrac{2a-3b}{4} + \dfrac{3a+2b}{5}$

Algebraic denominators

In the examples so far, the denominators have been numbers. If the denominators are algebraic, then we have to find an *algebraic* common denominator. This is a single algebraic expression which both denominators will divide into.

Suppose the expressions are $6a^2b$ and $15ab^3$. Take the (ordinary) least common multiple of 6 and 15, which is 30. For each of the variables, take the higher power. So we take a^2 and b^3. The common denominator is $30a^2b^3$.

Example   Write as a single fraction: $\dfrac{a}{6x^2} - \dfrac{b}{9xy}$.

The least common denominator of $6x^2$ and $9xy$ is $18x^2y$. Adjust both fractions.

$$18x^2y \div 6x^2 = 3y \qquad 18x^2y \div 9xy = 2x$$

So $6x^2$ must be multiplied by $3y$ to become $18x^2y$, and $9xy$ by $2x$.

$$\frac{a}{6x^2} - \frac{b}{9xy} = \frac{3y \times a}{3y \times 6x^2} - \frac{2x \times b}{2x \times 9xy} = \frac{3ya}{18x^2y} - \frac{2xb}{18x^2y}$$

$$\frac{a}{6x^2} - \frac{b}{9xy} = \frac{3ya - 2xb}{18x^2y}.$$

Exercise 16.8

Find the least common multiple of these expressions.

1 xy^2 and $7x^2y$

2 $10a^2b$ and $15ab^2$

3 $x+2$ and $x+3$

4 $x(x+2)$ and $x(x+3)$

Write the following as single fractions, simplifying your answers as far as possible.

5 $\dfrac{2}{x} + \dfrac{3}{y}$

6 $\dfrac{x}{a} - \dfrac{y}{b}$

7 $\dfrac{2}{x^2} + \dfrac{3}{x}$

8 $\dfrac{4}{xy} + \dfrac{3}{x^2}$

9 $\dfrac{5}{6xy^2} + \dfrac{3}{4x^2y}$

10 $\dfrac{2}{3a^2b} - \dfrac{5}{2ab^2}$

11 $\dfrac{2}{x+1} + \dfrac{3}{x+2}$

12 $\dfrac{3}{a-2} + \dfrac{4}{a+3}$

13 $\dfrac{4}{m+1} - \dfrac{3}{m-2}$

14 $\dfrac{1}{x} + \dfrac{3}{x^2} - \dfrac{1}{x+1}$

Equations with algebraic fractions

If an equation involves algebraic fractions, then the techniques described on pages 229–232 may have to be used.

Example Solve the equation $\dfrac{x+1}{2} + \dfrac{x}{3} = 8$.

Add the left-hand side.

$$\frac{x+1}{2} + \frac{x}{3} = \frac{3(x+1)+2x}{6} = \frac{3x+3+2x}{6} = \frac{5x+3}{6}$$

Hence $\dfrac{5x+3}{6} = 8$. Multiply by 6, and we get an ordinary linear equation.

$$5x + 3 = 48$$
$$5x = 45$$

The solution is $x = 9$.

Exercise 16.9

Solve these equations.

1 $\dfrac{x}{6} + \dfrac{x+2}{8} = 2$

2 $\dfrac{x+3}{4} + \dfrac{x-5}{3} = 9$

3 $\dfrac{x+2}{3} - \dfrac{x-3}{4} = 3$

4 $\dfrac{x-2}{2} - \dfrac{x-1}{3} = 3$

5 $\dfrac{x+5}{4} - \dfrac{x-4}{5} = 3$

SUMMARY

■ The **subject** of a **formula** is the letter written in terms of the other letters. For example, the subject of $y = mx + c$ is y.

■ When **changing the subject** of a formula, follow the same rules as for solving an equation.

■ **Algebraic fractions** can be **simplified** and combined in the same way as ordinary fractions.

For example, $\dfrac{ax}{ay} = \dfrac{x}{y}$ $\quad \dfrac{a}{b} \times \dfrac{x}{y} = \dfrac{ax}{by}$ $\quad \dfrac{a}{b} \div \dfrac{x}{y} = \dfrac{ay}{bx}$ $\quad \dfrac{a}{b} + \dfrac{x}{y} = \dfrac{ay + bx}{by}$

■ To solve an equation with algebraic fractions, use the rules for combining algebraic fractions.

Exercise 16A

1 Make x the subject of the equation $y = 6x + 5$.

2 Make a the subject of the equation $b = 4(a - 3)$.

3 Make r the subject of the formula $A = \pi k r^2$.

4 Make x the subject of the equation $mx + 5 = nx + 3$.

5 Make b the subject of the formula $a = \dfrac{b + 2}{b - 3}$.

6 Simplify the expression $\dfrac{14x^3}{21x^2 y}$.

7 Write $\dfrac{6x}{3y} \times \dfrac{15y}{2z}$ as a single fraction, simplifying your answer as far as possible.

8 Write $\dfrac{2x + 1}{3} + \dfrac{5x - 1}{2}$ as a single fraction, simplifying your answer as far as possible.

9 Write $\dfrac{2a}{y^2} - \dfrac{3b}{2xy}$ as a single fraction, simplifying your answer as far as possible.

10 Solve the equation $\dfrac{x + 3}{5} + \dfrac{x - 2}{3} = 9$.

Exercise 16B

1 Make a the subject of the equation $k = \frac{1}{2}a - 3$.

2 Make n the subject of the equation $m = 2(n + 8)$.

3 Make x the subject of the formula $y = \sqrt{x^2 - 3}$.

4 Make k the subject of the equation $ak - 1 = bk + 2$.

5 Make T the subject of the formula $S = \dfrac{T + a}{T + b}$.

6 Simplify the expression $\dfrac{x^2 + 2x}{xy + 2y}$.

7 Write $\dfrac{2}{x + 3} \div \dfrac{4}{x + 1}$ as a single fraction, simplifying your answer as far as possible.

8 Write $\dfrac{x + 4}{9} - \dfrac{3x + 4}{6}$ as a single fraction, simplifying your answer as far as possible.

9 Write $\dfrac{m}{2ab} - \dfrac{n}{6bc}$ as a single fraction, simplifying your answer as far as possible.

10 Solve the equation $\dfrac{x - 4}{4} - \dfrac{x + 2}{6} = 1$.

Exercise 16C (Ma1)

Which of the following are always true, and which are false? For the true ones, give a justification. For the false ones, give a counterexample.

1 $\dfrac{a}{b+c} = \dfrac{a}{b} + \dfrac{a}{c}$

2 $\dfrac{a+b}{c} = \dfrac{a}{c} + \dfrac{b}{c}$

3 $\dfrac{a+b}{c+d} = \dfrac{a}{c} + \dfrac{b}{d}$

4 $\dfrac{a}{a+b} = 1 - \dfrac{b}{a+b}$

5 $\dfrac{a-c}{b-c} = \dfrac{a}{b}$

Exercise 16D (Ma1)

Consider these five functions:

$$y = \frac{1}{x} \quad y = 1-x \quad y = 1-\frac{1}{x} \quad y = \frac{1}{1-x} \quad y = \frac{x}{x-1}$$

1 For each of these five functions, find y when $x = 3$.
2 Take each one of your answers to be the new value of x and repeat question 1. Do you notice anything? Test using one of your other answers to question 1.
3 Repeat questions 1 and 2 for another value of x. What do you notice?
4 Prove any rule that you have found.

The following exercise extends the material of this chapter to a topic you will meet if you do maths at A level.

Exercise 16E (Ma1)

In this chapter you have added together algebraic fractions. Often, it is necessary to go in the other direction, to write a complicated fraction in terms of simpler ones. In some cases these simpler fractions are called **partial fractions**.

1 Express as a single fraction: $\dfrac{4}{x+1} + \dfrac{6}{x+6}$.

2 Now try to go in the other direction. Find constants A and B such that

$$\frac{2x+7}{(x+1)(x+6)} = \frac{A}{x+1} + \frac{B}{x+6}$$

3 Find constants A and B such that

$$\frac{x+11}{(x+1)(x+6)} = \frac{A}{x+1} + \frac{B}{x+6}$$

4 Find constants A, B and C such that

$$\frac{x+2}{x^2(x+1)} = \frac{A}{x} + \frac{B}{x^2} + \frac{C}{x+1}$$

17

Three-dimensional geometry

To help you to think in terms of three dimensions, look around the room you are in. There are horizontal and vertical planes or faces (the floor and the walls) and there are horizontal and vertical lines or edges (two walls meet in a vertical line; the floor meets a wall in a horizontal line).

Exercise 17.1

1 The diagram shows a cube *ABCDEFGH*. The face *ABCD* is horizontal. Name
   a another horizontal face   b a vertical face
   c a horizontal line   d a vertical line.

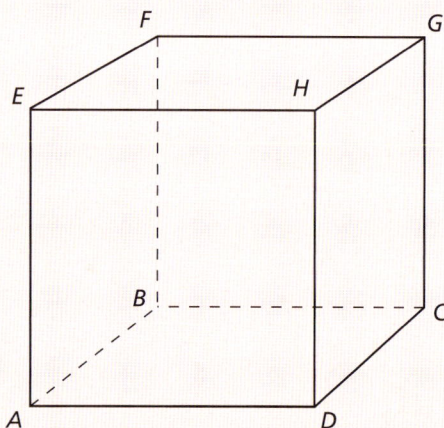

2 Which of the following are true?
  a Any two horizontal planes are parallel.
  b Any two vertical planes are parallel.
  c Any two horizontal lines are parallel.
  d Any two vertical lines are parallel.
  e Any two non-parallel planes meet in a line.
  f Any two non-parallel lines meet in a point.
3 Refer to the diagram for question 1. Write down
  a two faces which meet at *GC*   **b** the edges which meet at *F*.

A triangle is a flat, two-dimensional object. **Trigonometry** and **Pythagoras' theorem** are defined and used in two dimensions. But you can still use trigonometry and Pythagoras' theorem to find lengths and angles in solid, **three-dimensional** objects.

When you cut a solid, the flat surface exposed is a **section**. We speak of the cross-section of a solid – for example, the cross-section of a cylinder is a circle. In biology, a section is a cut which exposes part of a plant or an animal.

To find distances and angles in a solid, find a two-dimensional section of the solid, and apply trigonometry or Pythagoras' theorem in that section.

The difficulty is in finding which section to use, and in spotting where the right angles are. Here are some tips for identifying right angles in a three-dimensional shape.

- Any vertical line is perpendicular to any horizontal line.
- If a line is perpendicular to two different lines in a plane, then it is perpendicular to every other line in the plane.
- If a solid is drawn with one face in front of you, then any line leading away from the face is perpendicular to any line in the face.

Example The diagram shows a cube *ABCDEFGH*. Sketch the sections

  a *EGC*   **b** *FHA*

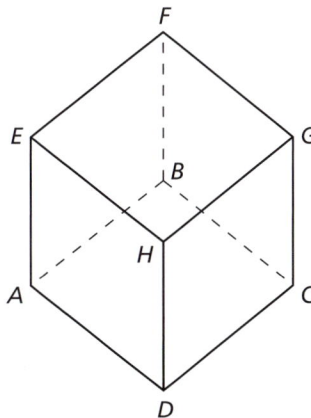

---

a *EG* is horizontal, and *GC* is vertical. Hence $\angle EGC = 90°$. *GC* is a side of one of the square faces, and *EG* is a diagonal of one of the square faces. Hence *EG* is longer than *GC*. The diagram shows the section.

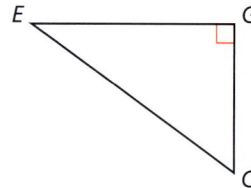

b *FH*, *HA* and *AF* are all diagonals of square faces. Hence they are equal in length. The section is an equilateral triangle, as shown.

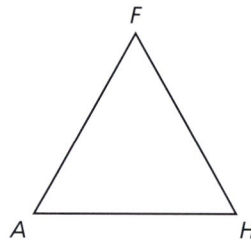

Exercise 17.2

1 The diagram shows a pyramid with a square base *ABCD*, and a vertex *V* above the centre *X* of the base.
  a Make a sketch of the section *VAB*. What sort of triangle is *VAB*?
  b Make a sketch of *VXA*. What sort of triangle is this?
  c Let the midpoints of *AB*, *BC* and *CD* be *L*, *M* and *N* respectively. Make sketches of *VLN* and *VLM*. What sort of triangle are they?

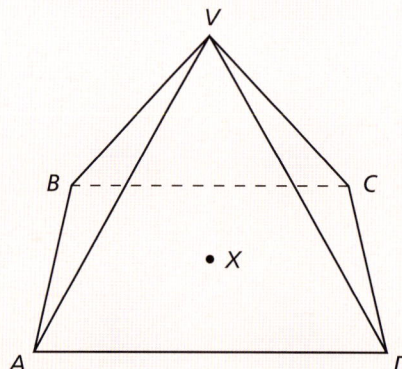

2 The diagram shows a regular tetrahedron $ABCD$. The midpoint of AB is M. Make a sketch of the section MCD.

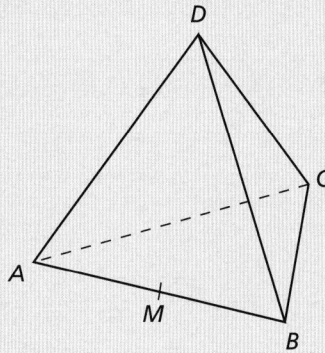

3 The diagram shows a regular octahedron $ABCDEF$. Make a sketch of $ECFA$. What sort of quadrilateral is it?

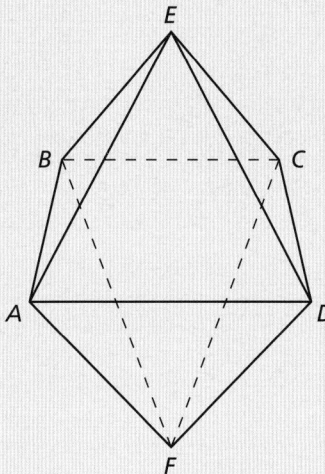

4 The diagram shows a prism whose cross-section is a right-angled triangle.
   a Make a sketch of $\triangle FBD$. What sort of triangle is it?
   b Make a sketch of $\triangle EBC$. What sort of triangle is it?

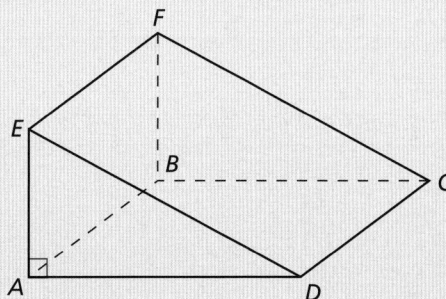

5 Look again at the cube in the example on page 237. Make a sketch of $FHDB$. What sort of quadrilateral is it?

Using Pythagoras' theorem

You can use Pythagoras' theorem to find lengths in three-dimensional shapes. But first you must find a two-dimensional section that is a right-angled triangle.

Example   The diagram shows a cube $ABCDEFGH$ of side 4 cm. Find the length of AG.

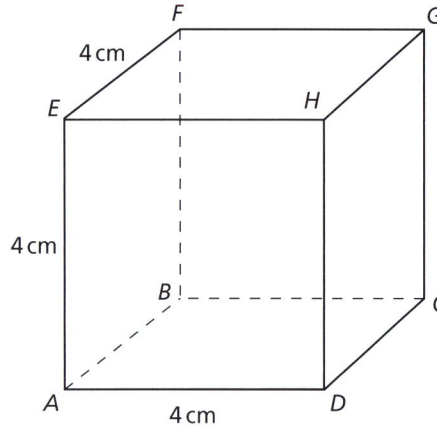

Consider the section AGC, as shown. This is a triangle with a right angle at C, as GC is vertical and AC is horizontal. We know that $GC = 4$ cm. So we need to find AC.

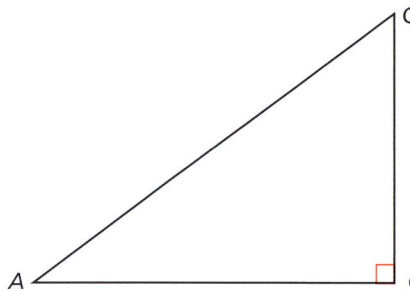

Consider the square $ABCD$. AC is a diagonal, so we can use Pythagoras' theorem in $\triangle ABC$. The length of AC is

$$\sqrt{AB^2 + BC^2} = \sqrt{4^2 + 4^2} = \sqrt{32}$$

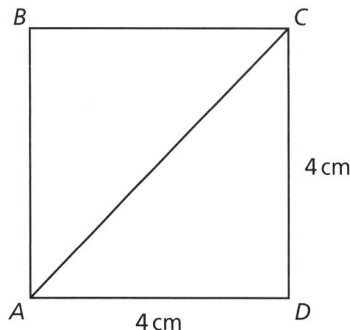

Now apply Pythagoras' theorem in $\triangle AGC$.

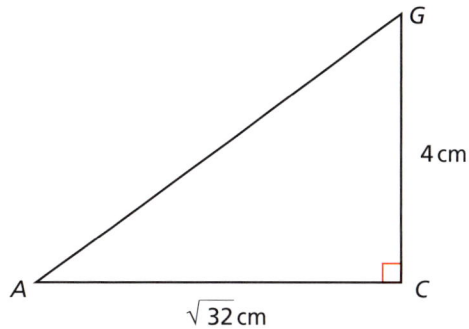

$$AG = \sqrt{AC^2 + GC^2} = \sqrt{(\sqrt{32})^2 + 4^2} = \sqrt{32 + 16} = \sqrt{48}$$

$AG = 6.93\,\text{cm}.$

Notes
1 We did not work out $\sqrt{32}$, as it was going to be squared later. To take the square root and then square the result runs the risk of introducing error.
2 Note that AG^2 is the sum of the squares of the three sides of the cube. That is

$$AG^2 = AB^2 + BC^2 + GC^2$$

> The greatest length is sometimes called the **space diagonal** of the cube.

This result is true for any cuboid. It is the three-dimensional version of Pythagoras' theorem. AG is the greatest length within the cube.

Exercise 17.3

1 $ABCDEFGH$ is a cuboid with $AB = 4\,\text{cm}$, $AD = 5\,\text{cm}$ and $AE = 6\,\text{cm}$. Find the length of AG.
2 In $ABCDEFGH$, the cube of side 4 cm in the example on page 240, let M be the midpoint of GH. Find the length of AM.
3 The diagram shows a pyramid $VABCD$, with $ABCD$ a square of side 2 m and $VA = VB = VC = VD = 3\,\text{m}$. The centre of $ABCD$ is X and the centre of AB is M. Find
   a XA **b** VM **c** VX

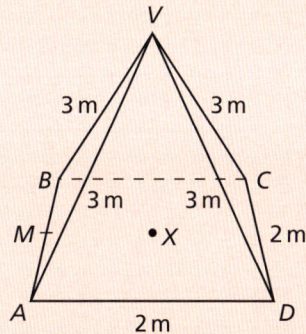

4 A square light-fitting of side 10 cm is suspended from its corners by four chains of length 30 cm, which are joined together at the ceiling. What is the depth of the square below the ceiling?

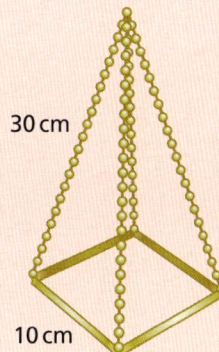

5 $VABCD$ is a pyramid, with $ABCD$ a rectangle. $AB = 10$ cm and $BC = 15$ cm. V is 12 cm above the centre X of $ABCD$. Find
a AC **b** AX **c** VA

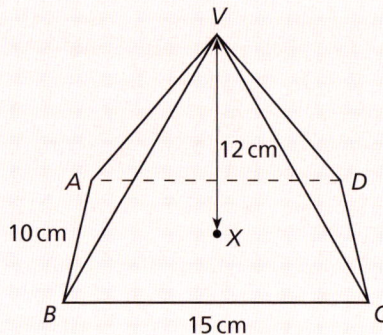

6 A cylinder has height 10 cm and base radius 6 cm. What is the greatest length within the cylinder?

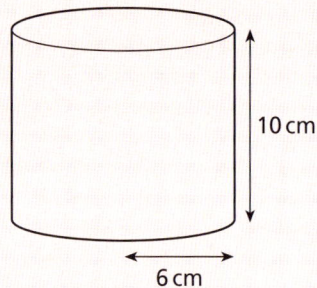

Angle between two lines

Two lines that meet are in a single plane. So we can use ordinary trigonometry to find angles and distances concerning the lines. In particular, we can find the angle between two lines.

Example $VABCD$ is a pyramid, with a square base of side 4 cm and $VA = VB = BC = VD = 5$ cm. Find the angle between VA and VC.

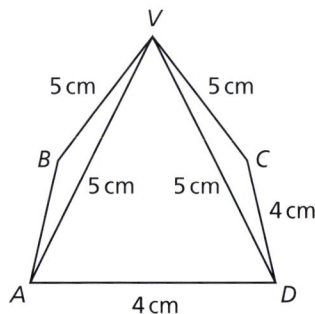

Take the section VAC. The angle we want is $\angle AVC$. The triangle is isosceles, with $VA = VC = 5$ cm, and AC is the diagonal of the base square. Hence

$$AC = \sqrt{AB^2 + BC^2} = \sqrt{4^2 + 4^2} = \sqrt{32}$$

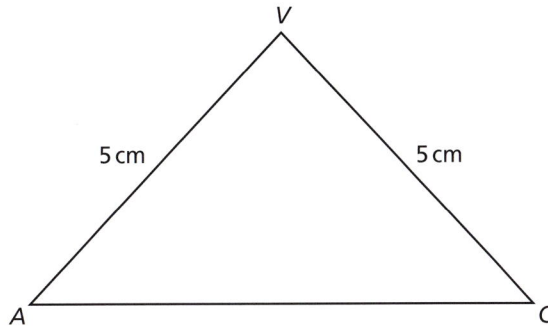

Let X be the midpoint of AC. Then $\angle VXA = 90°$, as VAB is an isosceles triangle.

$$\angle AVX = \sin^{-1}\left(\frac{AX}{VA}\right) = \sin^{-1}\left(\frac{\frac{1}{2}\sqrt{32}}{5}\right) = 34.4°$$

Double this to obtain $\angle AVC$.
   The angle between VA and VC is $68.9°$.

Exercise 17.4

1 $ABCDEFGH$ is a cube. Find the angle $\angle GBH$.

2 $ABCDEFGH$ is a cuboid, with $AB = 4$ cm, $AD = 5$ cm and $AE = 6$ cm. Find the angles
   a $\angle AFD$ **b** $\angle HBD$ **c** $\angle FDC$

3 The diagram shows a triangular prism $ABCDEF$. Find the angle between AC and AF.

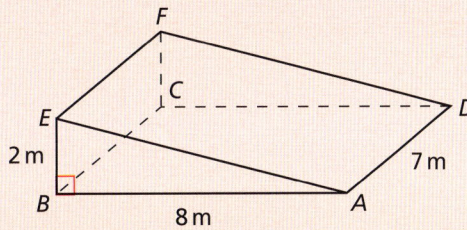

4 $ABCDEFGH$ is a cuboid, with $AB = 3$ cm, $AD = 4$ cm and $AE = 5$ cm. Find the angles
   a $\angle FBH$ **b** $\angle CEB$ **c** $\angle EHA$

5 The diagram shows a pyramid $VABCD$, with $ABCD$ a rectangle.
   $AB = 2$ m and $AD = 3$ m.
   $VA = VB = VC = VD = 4$ m.
   Find the angles
   a $\angle VAC$ **b** $\angle VAB$
   c $\angle DVC$ **d** $\angle DVB$

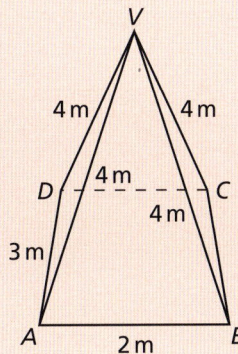

6 *VABCD* is a pyramid. *ABCD* is a square of side 10 m, and *V* is 16 m above the centre *X* of *ABCD*. Find the angles
  a $\angle VCD$ **b** $\angle VCA$ **c** $\angle DVB$ **d** $\angle DVC$

7 The diagram shows a regular octahedron. Find the angle $\angle EAC$.

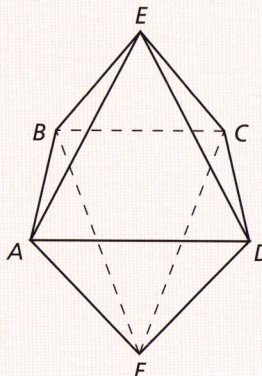

Angle between a line and a plane

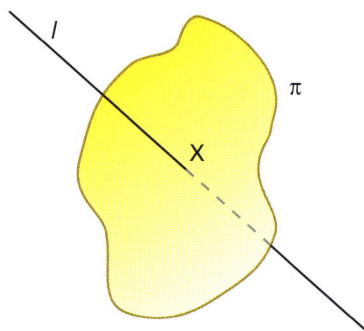

The diagram shows a line *l* passing through a plane π at *X*. What do we mean by the angle between *l* and π? The problem is that we could draw many lines in π, all making different angles with *l*, as shown.

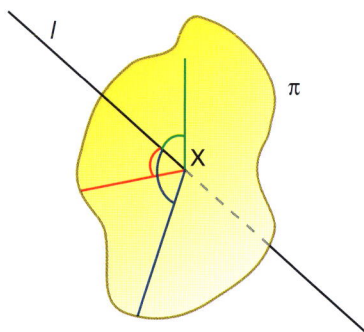